William Nanson Lettsom

The natural history of the tea-tree

with observations on the medical qualities of tea, and on the effects of

tea-drinking

William Nanson Lettsom

The natural history of the tea-tree
with observations on the medical qualities of tea, and on the effects of tea-drinking

ISBN/EAN: 9783741103995

Manufactured in Europe, USA, Canada, Australia, Japa

Cover: Foto ©Klaus-Uwe Gerhardt /pixelio.de

Manufactured and distributed by brebook publishing software
(www.brebook.com)

William Nanson Lettsom

The natural history of the tea-tree

Green Tea.

Fig.13.

Fig.16 *Fig.17*

Fig.12.

Fig.10 *Fig.11.*

P *Fig.15.*

S

P

S

Fig.14.

K

T *Fig.1.*

C

Fig.9.

K

C

Fig.4.

C

Fig.3.

K

Fig.2.

C

Fig.6.

C

Fig.7.

C

Fig.8.

Painted & Engrav'd by J. Miller.

Publish'd according to the Act of Parliament Dec. 10th 1771.

THE

NATURAL HISTORY

OF THE

TEA-TREE,

WITH OBSERVATIONS ON

THE MEDICAL QUALITIES OF TEA,

AND ON THE

EFFECTS OF TEA-DRINKING.

A NEW EDITION.

BY JOHN COAKLEY LETTSOM, M. D.

LONDON:

PRINTED BY J. NICHOLS;
FOR CHARLES DILLY.

1799.

ADVERTISEMENT.

IN the year 1769 was printed an inaugural differtation, in-tituled, " Obfervationes ad vires Theæ pertinentes."

In the year 1772 was publifhed, " The Natural Hiftory of " the Tea tree, with Obfervations on the Medical Qualities of " Tea, and Effects of Tea-drinking," which not only contained a tranflation of the Thefis, but likewife the natural hiftory of this vegetable, and which having been long out of print, it was thought a fecond edition would be favourably received by the publick.

In Sir George Staunton's Embaffy to China, lately publifhed, there are fome remarks on Tea, which are occafionally referred to in the prefent edition ; and they are referred to with the fatisfaction of confirming the relation firft offered to the publick in 1772.

As the Preface inferted at that time affords fome hints re-fpecting the introduction of the Tea-tree into Europe, it is pre-fixed to the prefent edition.

PREFACE

PREFACE

TO

THE FIRST EDITION, 1772.

THE subject of the following Essay being now in general use among the inhabitants of this kingdom, as well as in many other parts of Europe, and constituting a large part of our commerce, it cannot but afford pleasure to the curious to possess the history of a shrub, with the leaves of which they are so well acquainted.

Many treatises have been published on the uses and effects of Tea; a few writers have likewise given some circumstances relative to its natural history and preparation, the indefatigable Kæmpfer particularly; but these circumstances lie so dispersed, and the accounts which have been given of the virtues and efficacy of Tea are in general so contradictory, and void of true medical observation, that it still seemed no improper subject for a candid discussion. The reader may at least have the satisfaction of seeing, in a narrow compass, the principal opinions relative to this subject.

Within

Within thefe three or four years we have been fuccefsful enough to introduce into this kingdom a few genuine Tea plants. There was formerly, I am told, a very large one in England, the property of an Eaft-India captain, who kept it fome years, and refufed to part with either cuttings or layers. This died, and there was not another left in the kingdom. A large plant was not long fince in the poffeffion of the great Linnæus, but, I am informed, it is now dead. I know feveral gentlemen, who have fpared neither pains nor expence to procure this evergreen from China; but their beft endeavours have, in general, proved unfuccefsful. For, though many ftrong and good plants were fhipped at Canton, and all poffible care taken of them during the voyage, yet they foon grew fickly, and but one, till of late, furvived the paffage to England.

The largeft Tea plant in this kingdom is, I believe, at Kew; it was prefented to that royal feminary by John Ellis, Efq. who raifed it from the feed. But the plant at Sion-houfe, belonging to the Duke of Northumberland, is the firft that ever flowered in Europe; and an elegant drawing has been taken from it in that ftate, with its botanical defcription. The engraver has done juftice to his original drawing, which is now in the poffeffion of that great promoter of natural hiftory, Dr. Fothergill, to whom I have been indebted for many dried fpecimens and flowers of the Tea-tree from China. If the reader

compare

compare this plate with the following defcription, he will have as clear an idea of this exotic fhrub, as can at prefent be exhibited.

A few young Tea plants have lately been introduced into fome of the moft curious botanic gardens about London; hence it feems probable that this very diftinguifhed vegetable will become a denizen of England, and fuch of her colonies as may be deemed moft favourable to its propagation.

In regard to the effects of Tea on the human conftitution, one might have imagined that long and general ufe would have furnifhed fo many indifputable proofs of its good and bad properties, that nothing could be eafier than to determine thefe with precifion: yet fo difficult a thing is it to eftablifh phyfical certainty in regard to the operation of food or medicines on the human body, that our knowledge in general, even with refpect to this article, is very imperfect. Neverthelefs, I have endeavoured to avail myfelf of what has been written on this fubject by my predeceffors with the appearance of reafon, as well as of the converfation of learned and ingenious men now living, together with fuch experiments and obfervations as have occurred to me, fo as to furnifh the means of a more extenfive knowledge of the fubject.

With

With refpect to the prefent edition, fubfequent information has enabled me to enlarge it with fome important additions. Since the period of the original publication, the Tea-tree has been introduced into many of our gardens, and afforded the means of afcertaining its botanical characters. I have, at the fame time, the pleafure to obferve, that the firft edition has received the approbation of fome of the moft diftinguifhed botanifts. Linnæus, as well as Haller, as foon as they had perufed it, conveyed to me their approbation, in the kindeft manner : Murray and Cullen, and recently Schreber, have made frequent references to its authorities. If thefe diftinguifhed characters have approved the former, I am encouraged to hope that the prefent edition will not be lefs favourably received by the publick.

CONTENTS.

C O N T E N T S.

PART I.

THE NATURAL HISTORY OF THE TEA TREE.

PART II.

b   THE

THE

NATURAL HISTORY

OF THE

TEA-TREE.

PART THE FIRST.

SECTION I.
CLASS XIII. ORDER I.

POLYANDRIA MONOGYNIA.

K. CALYX,
Fig. 1, 2,
3. 10.

PERIANTHIUM
quinquepartitum,
minimum,
planum,
fegmentis *rotundis*,
obtufis,
perfiftentibus. (Fig.
1. K.)

K. The CA-
LYX, Fig.
1, 2, 3. 10.

A PERIANTHIUM
quinquepartite,
very fmall,
flat,
the fegments round,
obtufe,
permanent. (Fig. 1.
K.)

B

PETALA *fex*,	The PETALS fix[1],
fubrotunda,	fubrotund, or roundifh.
concava :	concave :
duo exteriora (F. 4. 7. C. C.)	two exterior, (F. 4. 7. C. C.)
minora,	lefs,
inæqualia,	unequal,

C. COROLLA, F. 1. 3, 4, 5, 6, 7, 8.

C. The Co-ROLLA, F. 1. 3, 4, 5, 6, 7, 8.

Nondum expanfa: (F. 3. C.)	the flower before it is fully blown: (F. 3. C.)
quatuor interiora, (F. 6. C. C. C. C. & F. 5.)	four interior (F. 6. C. C. C. C. and F. 5.)
magna,	large,
æqualia,	equal,
antequam decidunt, recurvata. (F. 8. C. C.)	before they fall off, recurvate. (F. 8. C. C.)

[1] Among feveral hundred fpecimens of dried Tea-flowers that I have examined, fcarcely one in twenty was perfect. Some had three petals only, fome nine, and others the feveral intermediate numbers. The greateft number confifted of fix large petals, and externally three leffer ones of the fame form. But the flowers, which bloffomed on the Tea-plant belonging to the duke of Northumberland, from which this defcription is taken, confifted in general of fix petals. One of the flowers indeed appeared to have eight petals ; however, the number in the flowers in moft plants vary confiderably, which may account for the miftake of Dr. Hill, and profeffor Linnæus (who defcribed this plant on Dr. Hill's authority), who make the green and bohea Tea two diftinct fpecies, giving nine petals to the former, and fix to the latter. See Amœn. Acad. Vol. VII. p. 248. Hill. Exot. t. 22. Kæmpfer. Amœn. Exot. p. 607. Breyn. Exot. Plant. Cent. 1. p. 111. Hift. de. l'Acad. des Sciences, 1776, p. 52.

STAMINA,

STAMINA, F. 6. 9, 10, 11.

f. FILAMENTA *nu-merofa*, (ducenta circiter.) (f. a. F. 6. 9.) *filiformia*, *corolla breviora*.

a. ANTHE RÆ cor-datæ, bi-loculares. (F. 10. 11.* Lente aucta.)

The STAMENS, F. 6. 9, 10, 11.

f. The FILAMENTS numerous [1], (f. a. Fig. 6. 9.) (about 200.) filiform, fhorter than the Corolla.

a. The AN-THERAS cordate, bi-locular [2]. (F. 10. 11.* magni-fied.)

PISTILLUM, F. 1. 10. 12.* Lente auctum.

g. GERMEN *globofo-trigonum.* (F. 1. 10. 12.)

s. STYLUS fimplex, ad apicem trifidus, (F. 12.)

Petalis Stamini-bufque delap-fis, a fe mu-tuo recedentes, divaricantes, &

The PISTIL-LUM, F. 1. 10. 12.* magnified.

g. The GERMEN three globular bodies joined (F. 1. 10. 12.)

s. The STYLE fimple, at the apex trifid, (F. 12.)

After the petals and ftamens are fallen off, they part from each other, fpread o-

[1] In a flower I received from that accurate naturalift, J. Ellis, F. R. S. &c. I counted upwards of 280 filaments; and, in another I had from Dr. Fothergill, there appeared to be nearly the fame number.
[2] Kæmpfer defcribes the Antheræ as being fingle.

PISTILIUM, F. 1. 10. 12 * Lente auctum.	longitudine aucta, marcefcentes. (F. 1. 12.) t. STIGMATA *fimplicia*. (F. 1. 9. 10. 12.)	The PISTIL-LUM, F. 1. 10. 12. * magnified.	pen, increafe in length, and wither on the Germen. (F. 1. 12.) t. The STIGMAS fimple. (F. 1. 9. 10. 12.)
P. PERICAR-PIUM, F. 1. 13, 14.	CAPSULA *ex tribus globis coalita*, (F. 13.) trilocularis, (F. 14.) apice trifariam dehifcens. (F. 13.)	P. The PE-RICARPI-UM, F. 1. 13. 14.	A CAPSULE in the form of three globular bodies united, (F. 13.) trilocular, (F. 14.) gaping at the top in three directions. (F. 13.)
S. SEMINA, F. 14.	*folitaria, globofa, introrfum angulata.*	S. The SEEDS, F. 14.	fingle, globofe, angular on the inward fide.
T. TRUNCUS, F. 1.	ramofus, lignofus, teres :	T. The TRUNK [1], F. 1.	ramofe, ligneous, round :

ramis

[1] Authors differ widely refpecting the fize of this tree. Le Compte fays, it grows of various fizes from two feet to two hundred, and fometimes fo thick, that two men can fcarcely grafp the trunk in their arms: though he afterwards obferves, that the Tea-trees, he faw in the province of Fokien, did not exceed five or fix feet in height.

	ramis alternis,		the branches alternate,
T. Truncus, F. 1.	vagis, rigidiufculis, cinerafcentibus, prope apicem rufefcentibus.	T. The Trunk, F. 1.	vague, *or placed in no regular order*, ftiffifh, inclining to an afh color, towards the top reddifh.
Pedunculi, (F. 1. p.)	axillares, (F. 1. p.) alterni, folitarii, curvati, uniflori, incraffati, (F. 1. 2. 7.) ftipulati : ftipula folitaria, fubulata, } (F. 1. 2. erecta. } 7. 9. d.)	Peduncles F. 1.	axillary, (F. 1. p.) alternate, fingle, curved, uniflorous, incraffate, (F. 1. 2. 7.) ¹ ftipulate : the ftipula fingle, fubulate, }(F. 1. 2. erect. } 7. 9. d.)

height. Journey through the empire of China. London, 1697, 8vo. p. 228. Du Halde quotes a Chinefe author, who defcribes the height of different Tea-trees, from one to thirty feet. Defcription génerale hiftorique, chronologique, politique, et phyfique de la Chine, Paris, 1755. Fol. 4 Tom. Hiftory of China, London, 1736. 8vo. Vol. IV. page 22. See alfo Guil. Pifo in Itinere Brafilica.

But Kæmpfer, who is chiefly to be depended upon, confines the full growth to about a man's heighth. Amœn. Exot. Lemgov. 1712, 4to. pag. 605. Probably this may be a juft medium ; for Ofbeck fays, that he faw Tea-fhrubs in flower-pots, not above an ell high. Voyage to China, Vol. I. p. 247. See alfo Eckeberg's account of the Chinefe hufbandry, Vol. II. p. 303.

¹ When the peduncles increafe in thicknefs towards their extremities.

alterna,

Pedunculi, (F. 1. p.)	alterna, elliptica, obtufe ferrata, marginibus inter dentes recurvatis.	Peduncles, P. 1.	alternate, elliptical, obtufely ferrate, edges between the teeth recurvate.

	apice marginata, (F. 15. e.) bafi integerrima, (F. 16. 17.) } * Lente aucta,		apex emarginate, (F. 15. e.) at the bafe very entire, (F. 16. 17.) } * magnified.
F. Folia, F. 1. 15, 16, 17.	glabra, nitida, bullata,	F. The Leaves, F. 1. 15, 16, 17.	fmooth, gloffy, bullate *,
	fubtus venofa,		venofe on the under fide,
	confiftentia, petiolata:		of a firm texture, on foot-ftalks:
	Petiolis breviffimis, (F. 1. 16. 17. b.)		The foot-ftalks very fhort, (F. 1. 16. 17. b.)

¹ No author has hitherto remarked this obvious circumftance; even Kæmpfer himfelf fays, that the leaves terminate in a fharp point. Amœn. Exot. p. 611.

² When the upper furface of the leaf rifes in feveral places in feveral places in roundifh fwellings, hollow underneath.

fubtus

F. Folia, F.
1. 15. 16.
17.
{
fubtus tereti-
bus,
}
(F. 16.
b.
* Lente
auctis.)

gibbis,

fupra plano - canali-
culatis. (F. 17.
b. * Lente auctis.)
}

F. The
LEAVES,
F. 1. 15,
16. 17.
{
round on the
under fide,
gibbous, *or*
bunching
out,
}
(F. 16.
b.
* mag-
nified.)

on the upper-fide,
flattifh, and flight-
ly channelled. (F.
17. b. * magni-
fied.)
}

Nomina   trivialia
Thea bohea &
viridis.

he common names
bohea and green
Teas'.

There is only one fpecies of this plant; the difference of
green and bohea Tea depending upon the nature of the foil,
the culture, and manner of drying the leaves. It has even
been obferved, that a green Tea-tree, planted in the bohea
country, will produce bohea Tea, and fo the contrary '.

' Whether the word TEA is borrowed from the Japanefe *Tijaa*, or the Chinefe
Theh, is not of much importance. By this name, with very little difference in pro-
nunciation, the plant here treated of is well known in moft parts of the world.
' I have examined feveral hundred flowers, both from the bohea and green Tea
countries, and their botanical characters have always appeared uniform. See Directions for bringing over feeds and plants from diftant countries, by John Ellis, Efq.
Sir George Staunton's Embafly, Vol. II. p. 464, fays, " Every information received
" concerning the Tea plant concurred in affirming that its qualities depended upon
" the foil in which it grew, and the age at which the leaves were plucked off the
" tree, as well as upon the management of them afterwards."

SECTION

SECTION II.

SYNONYMA.

MANY authors have at different times treated upon this
subject; some who never saw the Tea-tree, as well as others
who have seen it'. I shall enumerate those who are mentioned in the Species plantarum of Linnæus'.

Thea; Hortus Cliffort. 204. Mat. Med. 264. Hill. Exot.
t. 22.

Thee; Kæmpfer. Japan. 605. t. 606.

Thee frutex; Barthol. Act. 4. p. 1. t. 1. Bont. Jav. Amstel.
fol. 87 ad 88.

Thee Sinensium ; Breyn. Cent. 111. t. 112. incon. 17. t. 3.
Bocc. Muf. 114. t. 94.

Chaa; Casp. Bauhin. Pinax Theatri Botanici. Basil. 1623. 4to.
.147.

Evonymo affinis arbor orientalis nucifera, flore roseo; Pluk.
Alm. Botan. Stirp. nov. tradens. 1200. Lond. 1705. fol. 139.
t. 88. fig. 6.

In the Acta Haffnienfia, we meet with the first figure of this
tree; but, as it was taken from a dried specimen, it does not

---

' See Jac. Breynii Exotic. Cent. I. p. 114, 115.
' Vol. I. p. 734.

illuftrate

illuftrate the fubject very well. Bontius publifhed another, and though drawn in India, where he might have feen the plant, it does not much furpafs the preceding. The figure given by Plukenet is better than either of the former; and after his, Breynius publifhed one ftill better: but of all the engravings formerly executed, that given by Kæmpfer muft be allowed to be the moft accurate [1]; yet even this icon, like all the others publifhed by this induftrious naturalift, is extremely imperfect; although he certainly faw the living plants which he has reprefented, however expert the Chinefe may be in deception [2].

---

[1] Amœnit. Exotic. p. 618, et feq. See alfo his hiftory of Japan by Scheuchzer. Lond. 2 Vol. Fol. App. p. 3. Geoffr. Mat. Med. Vol. II. p. 276. Other figures of this fhrub are reprefented in Pifo Itinere Brafilico, Kircher's China Illuftrata, and Dutch Embaffy.

[2] Ofbeck, in his voyage to China, fpeaking of the Camellia, fays, " I bought one of a blind man in the ftreet, which had fine double white and red flowers. But, by farther obferving it in my room, I found that the flowers were taken from another; and one calyx was fo neatly fixed in the other with nails of bamboo, that I fhould fcarce have found it out, if the flowers had not begun to wither. The tree itfelf had only buds, but no open flowers. I learned from this inftance, that whoever will deal with the Chinefe, muft make ufe of his utmoft circumfpection, and even then muft run the rifk of being cheated." Vol. VII. p. 17. Mocquet in his Travels and Voyages, An. 1606, l. 4. p. 264, relates a curious piece of deception practifed by a Chinefe of Canton. " A Portuguefe," he fays, " bought a roafted duck at a cook's fhop in Canton. Seeing it look well, and appearing to be very fat, he carried it with him on-board his veffel, to eat it; but, when he had put his knife within it to cut it up, he found nothing but the fkin, which was upon fome paper, ingenioufly fitted up with little fticks, which made up the body of the duck; the Chinefe having very dexteroufly plucked away the flefh, and then fo well accommodated this fkin, that it feemed to be a true duck."

C   SECTION

SECTION III.

AUTHORS UPON TEA.

Besides the Authors already mentioned, feveral others have given fome account of this exotic ever-green, the principal of which are added for the farther information of thofe who may be defirous of confulting thefe writers on the fubject.

Johann. Petr. Maffeus rerum Indicarum libro vi. pag. 108. & lib. xii. pag. 242. Ludov. Almeyd. in eodem opere lib. iv. felect. epift.

Petr. Jarric. tom. III. lib. ii. cap. xvii.

Matth. Ric. de Chriftian. exped. apud Sinas, lib. i. cap. vii.

L. Baptifta Ramufio, le Navigationi e viaggi nelli quali fi Contienne la Defcrittione dell' Africa, del paefe del prete Joanni del mar Roffo, Calicut, ifole Moluchefe la Navigazione interno il mondo. Venet. 1550. 1563. 1588. 3 Vol. Fol. Vol. III. p. 15. -

Tranflation in Englifh of Giovanni Botaro¹, an eminent Italian author. Printed in 1590.

¹ This writer obferves, that the Chinefe have alfo an herb, out of which they prefs a delicate juice, which ferves them for a drink inftead of wine : it alfo preferves their health, and frees them from all thofe evils " that the immediate ufe of wine doth breed unto us." By the ufe the modern Chinefe make of Tea (who are a fober people) it can be nothing elfe. Anderfon's Chronolog. Deduction of Commerce.

Texeira,

Texeira, Relaciones del origen de los Reyes de Perſia y de
Hormuz. Amberes, 1610. p. 19.

Fiſcher's Sibiriſche Geſchichte, 1639. Vol. II. p. 694.

Alois Frois, in Relat. Japonicâ.

Nicol. Trigaut. de Regno Chinæ, Cap. 111. p. 34.

Linſcot. de Inſulâ Japonicâ, Cap. xxvi. p. 35. Ha. 1599.
Fol. et Belgiæ Amſt. 1644. Fol.

Bernhard. Varen. in deſcriptione Regni Japoniæ, Cap. xxiii.
p. 161.

Johan. Bauhin. Hiſtor. Univerſ. Plantarum, 1597. Tom. III.
lib. xxvii. cap. 1. p. 5. b.

Alex. Rhod. Sommaire des divers voyages et miſſions Apoſ-
toliques du R. P. Alexandre de Rhodes de la compagnie de Jeſus
à la Chine, et autres Royaumes de l'orient, avec ſon retour de
la Chine, à Rome; depuis l'année, 1618, juſques à l'annee,
1653, p. 25.

Dionyſii Joncquet, Stirpium aliquot paulò obſcurius officinis,
Arabibus aliiſque denominatarum, per Caſp. Bauhin. explicat.
p. 25. Ed. 1612.

Simon Pauli, Quadripartitum Botanicum, claſſe ſecundâ,
pag. 44. Ibidemque claſſe tertiâ, pag. 493.

Simon Pauli, Comment. de abuſu Tobaci et herbæ Theæ,
Roſtock. 1635. 4to. Straſburgh. 1665. Argent. 1665. 4to.
Francf. 1708. 4to. London, 1746. 8vo.

Wilhelm. Leyl. epiſtol. apud Simon Pauli in Comment. de
abuſu Tobaci, &c. p. 15. b.

Jacob. Bontii de Medicina Indorum, lib. iv. Leid. 1642.
12mo. et cum Piſone, Leid. 1658. Fol. Belgiæ, Ooſt en Weſt-
indiſche waarande, Amſtel. 1694. 8vo. Anglicè. An Account

of

of the Difeafes, Natural Hiftory, and Medicines, of the Eaft
Indies : London, 1769, 8vo.

Beginne ende voortgang van de Vereenighde Neederlande,
1646, 2 vol. et fub titulo, Recueil des Voyages faits pour
L'Etabliffement de la Campagnie des Indes Orientales, Amftel.
1702. 12mo. 10 Vol.

Joann. Nieuzofs, Gezantfchap an den Keizer van China,
p. 122. a.

, Erafmi Franciff. Oft-und Weft-Indifcher wie auch Sinefifcher
Luft-und Stats-Garten, p. 291.

Nicol. Tulpii, Obferv. Medic. lib. IV. cap. LX. p. 380.
Leidæ, 1641. 8vo.

Adam. Olearii, Perfionifche Reife-Befchreibung, 1633. p. 325.
lib. V. cap. XVII. p. 599. Fol. 1656. Hamburg. 1698. Amftel.
1666. 4to.

Johan. Albert. von Mandelflo, Morgenlandifche Reife-Befchreibung, lib. I. cap. XI. p. 39. Edit. 1656.

Olai Wormii, Muf. lib. II. cap. XIV. p. 165. Hafn. 1642. 4to.

Gulielm. Pifo, in Itinere Brafilico, Cliviæ, 1661. 8vo.

Athanaf. Kircher, Chin. Illuftrat. Ed. 1658. cum figura Fruct.
Theæ.

Simon de Molinariis, Ambrofia Afiatica, five de virtute et
ufu Theæ, Genuæ, 1672. 12mo.

De Comiers, le bon ufage du Thee, du Coffee, et du Chocolat, pour la Prefervation et pour la Guerifon des Malades, Paris,
1687. 12mo.

Marcus Mappus, de Thea, Coffea, et Chocolata. Argent. 1675
et 1695. 4to.

Oliv.

Oliv. Dappers, Befchryvinge des Keizerryts van Taifing or
Sina, Amftel. 1680. Fol. p. 226.

Nic. Blegny, du bon ufage du Thé, du Caffé, et du Chocolat.
Lyon. 1680. 12mo. Abrégé du traité du Caffé, &c. Lyon. 1687.
12mo.

John Overton, Voyage to Surat, London, 1696. 8vo.

John Overton, Effay upon the Nature and Qualities of Tea,
London, 1735. 8vo.

Paul Sylveftre du Four, de l'ufage du Thé, Caphè, et Chocolat. London, 1671. et auctius, 1684. 1686. 12mo. Hunc
libellum Jacobus Sponius Latinè reddidit, et edidit cum titulo,
Tr. nov. de potu Theæ, Coffeæ, Chocolatæ, Paris. 1685. 12mo.
cum figuris.

Pechlin, Theophilus Bibaculus, Franckfort, 1684. 4to.

Franc. Mich. Difdier, Befchreibung des Caffée, The, Chocolate, und Tobaks, Hamb. 1684. 12mo.

Bern. Albini, Difputatio de Thea, Francf. Viadr. 1684. 4to.

Arnold. Montan. Gudenfwaerdige Gefandtchappen aen de
Kaifaren van Japan. 1684.

J. Chamberlane, manner of making Tea, Coffee, and Chocolate, Lond. 1685. 12mo. p. 46.

Republiques des Lettres, tom. III. Fev. 1685.

Petri Petivi, Carmen de Thea; et Joh. Georg. Heinichen de
Theæ encomiis. Lugdun. 1685. 4to.

Corn. Bontekoe, van The, Coffy, en Chocolate. Haag. 1685.
8vo. Spanius de Thea, Coffea, et Chocolata.

Chriftian. Kurfner, de potu Theæ. Marpurg, 1681.

Jan.

Jan. Abraham. à Gehema, Weetftreit des Chinefifchen Thea
mit Warmen Waffer Berlin, 1685, 8vo. Francf. 1696. 8vo.
fub titulo, Zwanzig gefundheits regeln.

Steph. Blankaart gebrugk en mifbruyk van de Thee. Haag.
1686. 8vo.

The Natural Hiftory of Coffee, Tea, Chocolate, and Tobacco, with a Tract of the Elder and Juniper Berries, Lond.
1683. 4to.

Henrici Cofmii, magnæ naturæ œconomia cum demonftra-
tione Theæ, Coffeæ, Chocolatæ, Francf. Lipf. 1687. 12mo.

Elias Comerarius, in difputationibus de Thea et Coffea,
Tubingae, 1694. 8vo.

Le Compte's Journey through the Empire of China. Lond.
1697. 8vo. p. 228.

Joh. Ludov. Apinus, obf. 70. Decur. 3. Mifcell. Curiof.
1697. Andr. Cleyerus, Dec. 2. an. 4ti. pag. 7. Dan. Crugerus,
Dec. 2. Ann. 4ti. p. 141. Riedlinus Lin. Med. Ann. 4ti. Dom.
Ambrof. Stegmann, de Decoct. Theæ. Vol. V. p. 36.

Sir Thomas Pope Blount's Natural Hiftory, 8vo. . London,
1693.

Wilh. Ulrich Waldfchmidt, de ufu et abufu Theæ in genere.
Kiel. 1692. 8vo.

Ejufdem, an potus herbæ Theæ ecficcandi et emaciandi virtute pollerat ? Kiel, 1702. 4to.

P. Duncan, Avis Salutaire contre l'Abus du Coffe, du Chocolat, et du Thè. Rotterdam, 1705. 8vo. London, 1766.
8vo.

Groot mifbruyk van de Theæ en Coffæ. Haag, 1695. 4to.

Philofophical

Philofophical Tranfactions, V. I. an. 1665, 1666. Monday, July 2, 1766.

Plukenetii, Amalth. Botan. Londini, 1705, p. 79. 139.

Renaudot, anciennes relations de la Chine et des Indes. Paris, 1718, p. 31.

Kæmpfer, Amœnit. Exotic. Lemgov. 4to. 1712, p. 618. Les Lettres curieufes et edifiantes des Jefuites, paffim. Car. Frid. Luther, de potu Theæ, Kiel, 1712. 4to.

J. Cunningham, de variis fpeciebus Theæ, Agricultura Chinenfi, &c. n. 280.

Levuh. Frid. Meifner, Difputatio inaugur. de Thea, Coffea, Chocolata. Nuremb. 1721, 8vo.

Botanicum Officinale, or a compendious Herbal of fuch Plants as are ufed in Phyfic, by Jofeph Miller. Lond. 1722. 8vo.

Labat, Nouveau Voyage aux Iles de l'Amerique. Paris, 1721.

Joh. Henricus Cohaufon, Niewe Thee Tafel. et de Thea, Coffea, &c. à Chrift. Helwig. Amftel. 1719. 8vo. Germanicè, 1722. 8vo.

Short's Differtation upon the Nature and Properties of Tea, &c. London, 1730. 4to.

Ancient Accounts of India and China, by two Mahommedan Travellers. London, 1732.

L'Abbé Pluche, Le Spectacle de la Nature, à Paris, 1732.

Les Entretiens Phyfiques d'Arifte et d'Edoxe, par le pere Reynault. Paris, 1732. tom. 3.

John Arbuthnot, M. D. Effay concerning the Nature of Aliments. Lond. 1735. 8vo.

Cafp, Neumann, vom Thee, Coffee, Bier und Wein, Leipf. 1735.

J. Franc.

J. Franc. le Fevre, de natura, ufu, et abufu, Coffeæ, Theæ, Chocolatæ. Vefuntione, 1737. 4to.

R. James, Treatife on Tea, Tobacco, Coffee, and Choco-late, tranflated from Simon Pauli, Comment. &c. London, 1746, 8vo.

Barr. Rarior. 128. t. 904.

Du Halde, Defcription generale Hiftorique, Chronologique, Politique, et Phyfique, de la Chine, Paris, 1735. Fol. 4 vol. Haag. 1736. 4to. 4 vol. Hiftory of Japan, Lond. 1735. 8vo. 4 vol.

Aftley's Collection of Voyages, 4 vol. 4to. Lond. 1746.

The true Qualities of Tea. Anonymous. Lond. 1746. 8vo.

Petr. Kalms, Wäftgöta Refa, Stockholm, 1746. 8vo. tranf-lated by Forfter, Lond. 1772. 8vo. 2 vol.

James Stevenfon, Treatife on Tobacco, Tea, Coffee, and Chocolate, Lond. 1746. 8vo.

Chambers' Encyclopædia, Lond. 1752. Fol. Tom. II.

Mafon on the Properties of Tea, 1756. 8vo.

The good and bad Effects of Tea confidered, Anonymous, Lond. 1758. 8vo.

Linnæi Amœnit. Acad. V. vii. p. 241.

Newmann's Chemiftry, by Lewis. Lond. 1759. 4to. p. 373.

Hanway's Journal of eight Days Journey. London, 1759. 8vo. 2 vol. p. 21. vol. II.

Hanway's Obfervations on the Caufes of the Diffolutenefs amongft the Poor. Lond. 1772. 4to. p. 73. et paffim.

L'Abbé Jacquin, de la Santé utile à tout le Monde. à Paris, 1763. 8vo. p. 190.

Burmanni

Burmanni Fl. Indica, Lugd. Bat. 1766. p. 122.

Linnæi Sp. Plant. Vindobonæ. 1746. p. 734. Syft. Nat.
Vind. 1770. Tom. II. p. 365.

Linnæi Mat. Med. Vind. 1773. p. 136. Conf. Murray, appar. Med. Bergii Mat. Med. &c.

Encyclopedie, ou Dict. Raifonné, Neufch. 1765. Fol. Tom.
XVI. Thé.

M. de Begne de Prefle, de Confervateur de la Santé, ou Avis
fur les Dangers, &c. à Paris. 1763. 12mo. Dangers du Thé,
p. 118.

Concorde de la Geographie, ouvrage poftume de l'Abbé
Pluche, Paris, 1764. 12mo.

Will. Lewis, Experimental Hiftory of the Materia Medica,
Lond. 1768. 4to. p. 518.

Hart's Effays on Hufbandry. Lond. 1768. p. 166.

Tiffot on Difeafes incidental to literary and fedentary Perfons,
by Kirkpatrick. Lond. 1769. 12mo. p. 145.

Romaire Dictionaire d'Hiftoire naturelle. Paris, 1769. 8vo.

Milne's Botanical Dictionary, Lond. 1770. 8vo.

Brookes' Natural Hiftory. Lond. 1772. 6 vol. with a plate
of the Tea Plant.

Ofbeck's Voyage into China, by Forfter. Lond. 1771. 8vo.
2 vol.

Young's Farmer's Letters, Vol. I. p. 202. & 299.

Buc'hoz, Differtation fur le Thé fur la recolte, et fur les bons
et mauvais effets de fon infufian. Paris.

Blackwell's herbal. Lond. 1739. t. 351.

Thunberg, Flora Japon. Lipfiæ, 1784. p. 225.

D   Cullen's

Cullen's Mat. Med. Edinb. 1789. Tom. II. p. 309.

Murray, Appar. Medic. Gotting. 1787. Tom. IV. p. 226.

Grozier's general Defcription of China. London. 2 vol. 8vo. Vol. I. p. 442.

Fougeroux de Bondaroi, in Rozier, obf. et mem. fur la Phy-fique, Tom. I. f. 1.

Woodville's Supplement to Medical Botany. Lond. 1794. p. 116, with a figure.

Sir George Staunton, An authentic Account of an Embaffy, Lond. 1797. Vol. I. p. 22. and particularly Vol. II. p. 464.

SECTION

SECTION IV.

ORIGIN OF TEA.

As China and Japan [1] are the only countries known to us, where the Tea fhrub is cultivated for ufe, we may reafonably conclude, that it is indigenous to one of them, if not to both. What motive firft led the natives to ufe an infufion of Tea in the prefent manner is uncertain ; but probably in order to correct the water, which is faid to be brackifh and ill-tafted in many parts of thofe countries [2]. Of the good effects of Tea in fuch cafes, we have a remarkable proof in Kalm's journey through North America, which his tranflator gives us in the following words :

"Tea is differently efteemed by different people, and I think we *would* be as well, and our purfes much better, if we were without tea and coffee. However, I muft be impartial, and mention in praife of Tea, that if it be ufeful, it muft certainly be fo in fummer, on fuch journies as mine, through a

---

[1] Some authors add Siam alfo. Vid. Sim. Pauli Comment. et Wilh. Leyl. epift. apud Simon. Pauli comment. Nich. Tulpius obferv. Medicin. lib. IV. cap. lx. Lond. 1641.

[2] Le Compte's Journey through the Empire of China, p. 112. Staunton's Embaffy, Vol. II. p. 96. and particularly p. 68.

defart

defart country, where one cannot carry wine, or other liquors,
and where the water is generally unfit for ufe, as being full
of infects. In fuch cafes it is very pleafant when boiled, and
Tea is drank with it ; and I cannot fufficiently defcribe the fine
tafte it has in fuch circumftances. It relieves a weary traveller
more than can be imagined, as I have myfelf experienced, together with a great many others, who have travelled through
the defart forefts of America : on fuch journies Tea is found to
be almoft as neceffary as victuals [1]."

About the year 1600, Texeira, a Spaniard, faw the dried
Tea leaves in Malacca, where he was informed that the Chinefe
prepared a drink from this vegetable; and, in 1633, Olearius
found this practice prevalent among the Perfians, who procured the plant under the name of Cha orchia, from China, by
means of the Ufbeck Tartars. In 1639, Starkaw, the Ruffian
Ambaffador, at the Court of the Mogul, Chau Altyn, partook
of the infufion of Tea ; and, at his departure, was offered a
quantity of it, as a prefent for the Czar Michael Romanof,

[1] Kalm's Travels into North America, Vol. II. p. 314. The following note is
added by the ingenious Englifh tranflator in the 2d edition, Vol. II. p. 141 :

" On my travels through the defart plains, beyond the river Volga, I have had
feveral opportunities of making the fame obfervations on Tea ; and every traveller
in the fame circumftances will readily allow them to be very juft." Forfter, ibid.

See Brydone's Tour through Sicily and Malta, Let. 6. In letter 20, he fays, " We
have travelled all night on mules ; and arrived here about ten o'clock, overcome
with fleep and fatigue. We have juft had an excellent difh of tea, which never
fails to cure me of both ; and I am now as frefh as when we fet out." Captain
Forreft, in his Voyage to New Guinea, relates feveral inftances wherein the failors
experienced the exhilarating effects of this infufion.

which

which the Ambaſſador refuſed, as being an article for which he had no uſe [1].

This article was firſt introduced into Europe by the Dutch Eaſt India Company, very early in the laſt century; and a quantity of it was brought over from Holland about the year 1666 [2], by Lord Arlington and Lord Oſſory. In conſequence of this, Tea ſoon became known amongſt people of faſhion, and its uſe, by degrees, ſince that period, has become general.

It is, however, certain, that before this time, drinking Tea, even in public coffee-houſes, was not uncommon; for, in 1660, a duty of four-pence per gallon was laid on the liquor made and ſold in all coffee-houſes [3].

So

---

[1] Fiſcher's Libiriſche Geſchichte, Vol. II. p. 694—697. Monthly Magazine, Vol. VI. p. 60.

[2] Hanway's Journal of Eight Days Journey, Vol. II. p. 21. The ſame author obſerves, that Tea ſold at this time for ſixty ſhillings a pound. Anderſon, in his "Chronological Deduction of Commerce," remarks, that the firſt European author that mentions Tea wrote in the year 1590. However, by the preceding catalogue, it will appear, that this ſubject had been conſidered much earlier.

In Renaudot's anciennes Relations, Paris, 1718, p. 31, mention is made of two Arabian travellers who viſited China about the year 850; and related, that the inhabitants of that empire had a medicinal beverage, named chah or ſah, which was prepared by pouring boiling water on the dried leaves of a certain herb, which infuſion was reckoned an efficacious remedy in various diſeaſes.

[3] By an act made this year, the duties of Exciſe on malt liquor, cyder, perry, mead, ſpirits, or ſtrong waters, coffee, tea, ſherbet, and chocolate, were ſettled on the King during his life. Then it was that Coffee, Tea, and Chocolate, were firſt mentioned in the ſtatute book. Noorthouck, in his Hiſtory of. London, remarks, that King Charles II. iſſued a proclamation for ſhutting up the coffee-houſes, &c. about a month after he had dined with the Corporation of London, at Guildhall, on their Lord-Mayor's day, Oct. 29, 1675. At this feaſt the King afforded

So early as 1678, Cornelius Bontekoe, a Dutch phyfician, publifhed a treatife, in his own language, on Tea, Coffee, and Chocolate '. In this he fhews himfelf a very zealous advocate for Tea, and denies the poffibility of its injuring the ftomach, although taken to the greateft excefs, as far as one or two hundred cups in a day. To what motive we are to impute the partiality of Dr. Bontekoe, is uncertain at this period; but as he was firft phyfician to the Elector of Brandenburgh, and probably of confiderable eminence and character, his eulogium might

forded the Citizens abundant matter for animadverfion, in which they indulged themfelves fo much to his diffatisfaction, and that of his *cabal* miniftry, that a proclamation was iffued December 20, for fhutting up and fuppreffing all coffee-houfes; " becaufe, in fuch houfes, and by occafion of the meeting of difaffected perfons in them, divers falfe, malicious, and fcandalous reports were devifed and read abroad, to the defamation of his Majefty's government, and to the difturbance of the quiet and peace of the realm." The opinions of the judges were taken on this great point of ftopping people's tongues, when they fagely refolved, " that retailing of Coffee and Tea might be an innocent trade; but as it was ufed to nourifh fedition, fpread lies, and fcandalize great men, it might alfo be a common nuifance." In fhort, on a petition of the merchants and retailers of Coffee and Tea, permiffion was granted to keep open the coffee-houfes until the 24th of June next, under an admonition, that the mafters of them fhould prevent all fcandalous papers, books, and libels, from being read in them; and hinder every perfon from declaring, uttering, or divulging all manner of falfe or fcandalous reports againft government or the minifters thereof. Thus, by a refinement of policy, the fimple manufacturer of a difh of Coffee or Tea was conftituted licenfer of books, corrector of manners, and arbiter of the truth or falfehood of political intelligence over every company he entertained! And here the matter ended. Chap. 15.

In May 1784 an act was paffed, called the Commutation Act, " for repealing the feveral duties on Tea, and for granting to his Majefty other duties in lieu thereof; and alfo feveral duties on inhabited houfes."

' The fecond edition was publifhed under the title of Van The, Coffy, en Chocolate. Haag. 1685. 8vo. The late Baron Van Swieten cenfures this phyfician for his remarkable bias in favour of this exotic. Comment. Vol. V. p. 587. Eft modus in rebus, may be as aptly applied to Dr. Bontekoe as to Dr. Duncan.

tend

tend greatly to promote its ufe : however, we find its importation and confumption were daily augmented ; and, before the conclufion of the laft century, it became generally known among the common people in England.

It is foreign to my fubject, or it would perhaps afford to a fpeculative mind no inconfiderable fatisfaction, to trace the confumption from its firft entrance at the Cuftom-houfe to the prefent amazing imports. At this time upwards of twenty-three millions of pounds are annually allowed for home confumption ; and the Eaft India Company have generally in their warehoufes a fupply at leaft for one year.

The following account of the importation of Tea, from 1776 to 1795, as related by Sir George Staunton (Vol. II. p. 624), may be fatisfactory to the Reader :

An Account of the Quantities of Teas exported from China, in English and Foreign Ships, in each Year from 1776 to 1795, distinguishing each Year.

	1776.	ships	1777.	ships	1778.	ships	1779.	ships	1780.	ships	1781.	ships	1782.	ships	1783.	ships	1784.	ships	1785.	ships
By Swedes	lb. 2,652,500	2	3,049,100	2	2,851,100	2	3,338,900	2	2,626,900	2	4,108,900	3	3,267,000	3	4,265,600	3	4,898,000	4	8,158,000	4
Danes	2,833,700	2	2,487,300	2	2,098,300	2	3,388,400	3	3,983,600	3	3,344,400	3	4,118,500	3	5,477,200	4	4,222,000	4	5,334,000	4
Dutch	4,933,700	3	4,836,500	4	4,695,100	4	4,551,100	4	4,687,800	4	4,957,600	4							4,960,000	4
French	2,531,600	5	5,719,100	7	3,675,500	4	2,102,800	4	1,375,600	1							4,231,200	4	
Imperial											217,700	1					3,484,200	5	
Hungarian															933,300				3,199,000	4
Tuscan															3,954,100	8			880,100	2
Portugueze																				
American																				
Prussia																				
Spanish																			3,339,800	
Total Foreign	lb. 12,841,500	13	16,112,000	15	13,002,700	11	11,302,300	11	12,673,700	10	11,725,600	5	7,385,800	16	14,630,200	18	19,072,200	21	17,531,100	18
English private Trade included	3,492,415	8	5,673,434	9	6,392,788	7	4,373,021		none imported		11,592,819	9	6,857,731	6	4,138,295	13	9,016,760	14	10,583,628	14
	lb. 16,243,915	21	21,785,434	24	19,695,488	18	15,674,321	10	12,673,700	10	23,318,419	14	14,243,531	22	18,768,495	14	28,089,060	33	28,114,728	32

	1786.	ships	1787.	ships	1788.	ships	1789.	ships	1790.	ships	1791.	ships	1792.	ships	1793.	ships	1794.	ships	1795.	ships
By Swedes	lb. 6,212,400	4	1,747,700	2	2,890,000	2	2,589,000	2					1,591,330	1	1,359,730	1	256,130		
Danes	4,578,100	3	2,093,000	2	2,064,000	2	2,496,800	2	2,773,000	1	520,700		1,591,330	1	852,260		24,670	1
Dutch	4,456,500	5	5,943,200	5	5,724,000	5	4,170,600	5	1,388,500	3	2,051,330	2	2,938,530	2	4,056,800	4
French	406,000	1	382,266	3	1,728,000	1	293,100	1	3,943,100	2	443,105	4	784,000	2	1,540,070	2	
Imperial																			
Hungarian																			
Tuscan																			
Portugueze	695,000	5	1,181,866	2	750,000	4	1,188,800	14	5,093,200				1,869,200	6	393,876		1,438,270	7
American					409,300	2	818,400	2			743,100	1	5,070		1,538,400	7	1,074,130	
Prussia																	400		289,470	1
Genoese											260				578,930	3	17,460	1
Total Foreign	lb. 16,110,920	13	11,347,000	14	14,328,000	15	1,064,700	15	10,267,400	21	3,024,660	10	6,294,930	19	9,423,000	12	3,436,930	14	5,577,200	14
English private Trade included	13,480,691	27	20,610,919	29	23,096,795	27	20,141,745	31	17,991,033	31	22,369,610	35	13,185,467	16	26,005,414	18	20,728,795	21	23,733,810	
	lb. 30,591,591	44	31,957,939	44	36,424,605	42	31,206,445	42	28,258,432	41	25,424,282	33	19,480,397	35	35,428,614	30	26,165,635	35	29,311,010	

* Most of these foreign ships went to China, previous to the Commutation Act, which passed into effect in England in September, 1784.
‡ Part of these should have arrived in 1780.

It is probable that the Dutch, as they traded confiderably to Japan about the time Tea was introduced into Europe, firft brought this article from thence. But now China is the general mart, and the province Fokien, or Fo-chen ', the principal country, that fupplies both the Empire and Europe with this commodity.

' In this province, this fhrub is called Thee, or Te ; and as the Europeans firft landed here, that dialect has been preferred. Le Compte's Journey through the Empire of China, p. 227. Du Halde's Hiftory of China, Vol. IV. p. 21.

E   SECTION

SECTION V.

SOIL AND CULTURE.

To the ingenious Kæmpfer we are principally indebted for any accurate information refpecting the culture of the Tea Tree; and, as his account was compofed during his refidence at Japan, greater credit is certainly due to it. We fhall give what he fays upon this fubject, and then ftate the accounts we have been able to collect of the Chinefe method.

Kæmpfer tells us, that no particular gardens or fields are allotted for this plant, but that it is cultivated round the borders of rice and corn fields, without any regard to the foil. Any number of the feeds, as they are contained in their feed veffels, not ufually lefs than fix, or exceeding twelve or fifteen, are promifcuoufly put into one hole, made four or five inches deep in the ground, at certain diftances from each other. The feeds contain a large proportion of oil, which is foon liable to turn rancid; hence fcarce a fifth part of them germinate, and this makes it neceffary to plant fo many together.

The feeds vegetate without any other care; but the more induftrious annually remove the weeds, and manure the land. The leaves which fucceed are not fit to be plucked before the

third

third year's growth, at which period they are plentiful, and in their prime.

In about feven years the fhrub rifes to a man's height ; but as it then bears few leaves, and grows flowly, it is cut down to the ftem, which occafions fuch an exuberance of frefh fhoots and leaves the fucceeding fummer, as abundantly compenfates the owners for their former lofs and trouble. Some defer cutting them till they are of ten years growth.

So far as can be gathered from authors and travellers of credit, this fhrub is cultivated and prepared in China, in a fimilar manner to what is practifed in Japan ; but as the Chinefe export confiderable quantities of Tea, they plant whole fields with it, to fupply foreign markets, as well as for home confumption.

The Tea-tree delights particularly in vallies ; or on the declivities of hills, and upon the banks of rivers, where it enjoys a fouthern expofure to the fun ; though it endures confiderable variations of heat and cold, as it flourifhes in the northern clime of Pekin, as well as about Canton [1], the former of

---

[1] The beft Tea grows in a mild temperate climate ; the country about Nankin producing better Tea than either Pekin or Canton, between which places it is fituated. It has been afferted, that no Tea-plants have yet died in England through excefs of cold ; but the contrary, I know, has happened. The plant in the Princefs Dowager's garden at Kew flourifhed under glafs windows, with the natural heat of the fun, as now do thofe at Mile-end, in the poffeffion of the intelligent Botanift J. Gordon. The Tea-plant belonging to Dr. Fothergill thrives in his garden at Upton, expofed to the open air, and the plant introduced into the Botanic garden at Chelfea had one leaf which meafured five inches and a quarter in length.

which

which is in the fame latitude with Rome; and from meteo-
rological obfervations it appears, that the degree of cold about
Pekin is as fevere in winter, as in fome of the northern parts.
of Europe '.

' Du Halde and other authors have obferved, that the degree of cold in fome
parts of China is very fevere in winter. In the inland parts of North America, and
on extenfive continents, the degrees of heat and cold are found to be much more
violent than in iflands or places bordering on the fea in the fame latitude, as the
air that blows over the fea is liable to lefs variation in thefe refpects, than that
which blows over large tracts of land.

SECTION VI.

GATHERING THE LEAVES.

A t the proper feafons for gathering the Tea leaves, labourers are hired, who are very quick in plucking them, being accuſtomed to follow this employment as a means of their livelihood. They do not pluck them by handfuls, but carefully one by one ; and, tedious as this may appear, each perſon is able to collect from four to ten or fifteen pounds in one day. The different periods in which the leaves are uſually gathered, are particularly deſcribed by Kæmpfer [1].

I. The firſt commences at the middle of the laſt moon, immediately preceding the vernal equinox, which is the firſt month of the Japaneſe year, and falls about the latter end of our February, or beginning of March. The leaves collected at this time are called Ficki Tsjaa, or powdered Tea, becauſe they are pulveriſed and ſipped in hot water (Sect. IX. 1). Theſe tender young leaves are but a few days old when they are plucked ; and, becauſe of their ſcarcity and price, are diſpoſed of to princes and rich people only ; and hence this kind is called Imperial Tea.

[1] Amœnitat. Exotic. p. 618, et ſeq. Hiſtory of Japan. Appendix to Vol. II. p. 6, et ſeq.

A ſimilar

A fimilar fort is alfo called Udfi Tsjaa, and Tacke Sacki Tsjaa, from the particular places where it grows. The peculiar care and nicety obferved in gathering the Tea leaves in thefe places deferve to be noticed here, and we fhall therefore give fome account of one of them.

Udfi is a fmall Japanefe town, bordering on the fea, and not far diftant from the city of Miaco. In the diftrict of this little town, is a pleafant mountain of the fame name, which is thought to poffefs the moft favourable foil and climate for the culture of Tea, on which account it is inclofed with hedges, and likewife furrounded with a broad ditch for farther fecurity. The trees are planted upon this mountain in fuch a manner as to form regular rows, with intervening walks. Perfons are appointed to fuperintend the place, and preferve the leaves from injury or dirt. The labourers who are to gather them, for fome weeks before they begin, abftain from every kind of grofs food, or whatever might endanger communicating any ill flavour to the leaves; they pluck them alfo with the fame delicacy, having on a thin pair of gloves[1]. This fort of imperial or bloom Tea[2] is afterwards prepared, and then efcorted by the chief furveyor of the works of this mountain, with a ftrong guard, and a numerous retinue, to the emperor's court, for the ufe of the Imperial family.

[1] The fame cautions are not ufed previous to collecting other forts of Tea.

[2] This cannot be the fort to which alfo the Dutch give that name, as it is fold upon the fpot to the princes of the country, for much more than the common bloom Tea is fold for in Europe. Kæmpfer, Amœnit. Exotic. p. 617. Hiftory of Japan, Appendix, p. 9. Neumann's Chemiftry by Lewis, p. 373.

II. The

II. The fecond gathering is made in the fecond Japanefe month, about the latter end of March, or beginning of April. Some of the leaves at this period are come to perfection, others not arrived at their full growth; both however are promifcuoufly gathered, and are afterwards forted into different claffes, according to their age, fize, and quality; the youngeft particularly are carefully feparated, and are often fold for the firft gathering or Imperial Tea. The tea collected at this time is called Tootsjaa, or Chinefe Tea, becaufe it is infufed, and drank after the Chinefc manner (SECT. IX. 1.) It is divided by the Tea-dealers and merchants into four kinds, diftinguifhed by as many names.

III. The third and laft gathering is made in the third Japanefe month, which falls about our June, when the leaves are very plentiful and full grown. This kind of Tea, called Ban Tsjaa, is the coarfeft, and is chiefly drank by the lower clafs of people (SECT. IX. 111.)

Some confine themfelves to two gatherings in the year, their firft and fecond anfwering the preceding fecond and third. Others have only one general gathering [1], which they make alfo at the fame time with the preceding third or laft gathering: however, the leaves collected at each time, are refpectively feparated into different fortments.

The Chinefe collect the Tea at certain feafons [2], but whether the fame as in Japan, we are not fo well informed, moft pro-

---

[1] In this cafe the under leaves, which are harfh and lefs fucculent, are probably left upon the trees. See Eckeberg's Chinefe Hufbandry in Ofbeck's Voyage, Vol. II. p. 303.

[2] Du Halde's Hiftory of China, Vol. IV. p. 21.

bably

bably, however, the Tea harveſt is nearly at the ſame periods, as the natives have frequent intercourſe, and their commercial concerns with each other are very extenſive [1].

[1] Du Halde, Vol. II. p. 300. Kæmpfer obſerves, in his Hiſtory of Japan, that the trade between theſe nations has continued from remoteſt antiquity; formerly the Chineſe had a much more general intercourſe with the Japaneſe than they have at preſent; the affinity in the religion, cuſtoms, books, learned languages, arts and ſciences of the Chineſe with the latter, procured them a free toleration in Japan. Hiſtory of Japan, Vol. I. p. 374.

SECTION

SECTION VII.

METHOD OF CURING OR PREPARING TEA
IN JAPAN.

Public buildings, or drying houfes, are erected for curing Tea, and fo regulated, that every perfon, who either has not fuitable conveniences, or wants the requifite fkill, may bring his leaves at any time to be dried. Thefe buildings contain from five to ten or twenty fmall furnaces, about three feet high, each having at the top a large flat iron pan [1], either high, fquare, or round, bent up a little on that fide which is over the mouth of the furnace, which at once fecures the operator from the heat of the furnace, and prevents the leaves from falling off.

There is alfo a long low table covered with matts, on which the leaves are laid, and rolled by workmen, who fit round it. The iron pan being heated to a certain degree by a little fire made in the furnace underneath, a few pounds of the frefh-gathered leaves are put upon the pan ; the frefh and juicy leaves crack when they touch the pan, and it is the bufinefs of the

---

[1] Some writers mention copper pans, and fuppofe that the green efflorefcence which appears on copper may increafe the verdure of green Tea; but, from experiments that I made, there does not appear any foundation for this fuppofition. See SECT. VIII.

F   operator

operator to fhift them as quick as poffible with his bare hands, till they grow too hot to be eafily endured. At this inftant he takes off the leaves, with a kind of fhovel refembling a fan, and pours them on the matts to the rollers, who, taking fmall quantities at a time, roll them in the palms of their hands in one direction, while others are fanning them, that they may cool the more fpeedily, and retain their curl the longer[1].

This procefs is repeated two or three times, or oftener, before the Tea is put in the ftores, in order that all the moifture of the leaves may be thoroughly diffipated, and their curl more completely preferved. On every repetition the pan is lefs heated, and the operation performed more flowly and cautioufly[2]. The Tea is then feparated into the different kinds, and depofited in the ftore for domeftic ufe or exportation.

As the leaves of the Ficki Tea (Sect. VI. and IX. 11.), are ufually reduced into a powder before they are drank, they fhould be roafted to a greater degree of drynefs. As fome of thefe are gathered when very young, tender, and fmall, they are firft immerfed in hot water, taken out immediately, and dried without being rolled at all.

Country people cure their leaves in earthen kettles[3], which anfwer every neceffary purpofe at lefs trouble and expence, whereby they are enabled to fell them cheaper.

[1] Sir G. Staunton, Embaffy to China, obferves that the Tea leaves are each rolled feparately between the fingers of a female, Vol. II. p. 465.

[2] This fhould be carefully attended to, in curing the fine green Teas, to preferve their verdure and perifhable flavour. See Sect. VIII. ad finem.

[3] This is alfo done in China. See Eckeberg's Chinefe Hufbandry in Ofbeck's Voyage, Vol. II. p. 303.

To

To complete the preparation, after the Tea has been kept for fome months, it muft be taken out of the veffels, in which it had been contained, and dried again over a very gentle fire, that it may be deprived of any humidity which remained, or might fince have been contracted.

The common Tea is kept in earthern pots with narrow mouths; but the beft fort of Tea ufed by the Emperor and nobility is put in porcellane or China veffels. The Bantsjaa, or coarfeft Tea, is kept by the country people in ftraw bafkets, made in the fhape of barrels, which they place under the roofs of their houfes, near the hole that lets out the fmoke, and imagine that this fituation does not injure the Tea.

This is the relation we have from Kæmpfer of the method in which the Japanefe collected and cured their Tea. In the accounts of China, authors have in general treated very flightly of the cultivation and preparation of Tea. Le Compte[1] indeed obferves, that to have good Tea, the leaves fhould be gathered while they are fmall, tender, and juicy. They begin commonly to gather them in the months of March and April, according as the feafon is forward; they afterwards expofe them to the fteam of boiling water to foften them; and, as foon as they are penetrated by it, they draw them over copper plates[2]

---

[1] Journey through the Empire of China.

[2] Upon this fubject, fee Sect. VII. and VIII. It may be doubted alfo whether the conclufion of Le Compte's relation is not erroneous, as it is improbable that any leaves fhould of themfelves take fo perfect a curl as that in which Tea is brought into Europe. No materials are ufed but iron and earthen for drying Tea, as obferved in note ', p. 33.

kept

kept on the fire, which dries them by degrees, till they grow brown, and roll up of themselves in that manner we fee them.

However, it is certain, from the Chinese drawings, which exhibit a faithful picture, though rudely executed, of the whole procefs from beginning to end, that the Tea tree grows for the moft part in hilly countries, on their rocky fummits, and fteep declivities; and it would feem by the pains the Chinefe are at, in making paths, and fixing a kind of fcaffolds, to aflift them, that thefe places afford the fineft Tea. It appears from thefe drawings, that the trees in general are not much taller than man's height: The gatherers of the leaves are never reprefented but on the ground; they make ufe of hooked fticks indeed, but thefe feem rather intended to draw the branches towards them, when they hang over brooks, rivers, or from places difficult of accefs, than to bend down the tops or upper branches of the trees on plain ground.

They pick the leaves as foon as gathered into different forts, and cure them nearly in the manner defcribed to be practifed by the Japanefe. They build a range of ftoves, like thofe in a chemift's laboratory, or great kitchen, where the men work, and curl the leaves in the pans themfelves. It feems alfo that they repeat the drying. They dry it likewife, after having fpread it abroad in fhallow bafkets, in the fun; and, by the means of fieves, feparate the larger from the fmaller leaves, and thefe again from the duft.

The Chinefe put the finer kinds of Tea into conic veffels, like fugar loaves, made of tutenaque, tin, or lead, covered

with

with neat matting of bamboo ; or in fquare wooden boxes lined
with thin lead, dry leaves and paper, in which manner it is exported to foreign countries. The common Tea is put into
bafkets, out of which it is emptied, and packed up in boxes or
chefts as foon as it is fold to the Europeans [1].
One thing fhould be mentioned to their credit ; when their
harveft of Tea is finifhed, each family fails not to teftify, by
fome religious rite, their gratitude to the Giver.

[1] There are feveral difgufting circumftances attending the preparation of Tea.
Ofbeck fays, the Chinefe fervants tread the Tea into the chefts with their naked
feet. Voyage to China, Vol. I. p. 252. Sir George Staunton makes a fimilar re-
mark, Vol. II. p. 466.

SECTION

SECTION VIII.

VARIETIES OF TEA.

I⊤ has been already obferved (SECT. VI.) that many different fortments of Tea are made during the times of collecting the leaves; and thefe are multiplied according to the goodnefs of their preparation, by which the varieties of Tea may be confiderably augmented '. The diftinctions with us are much more limited, being generally confined to three principal kinds of green, and five of bohea.

I. Thofe of the former are,

1. Bing, imperial, or bloom Tea, with a large loofe leaf, of a light green colour, and faint delicate fmell.

11. Hy-tiann, hi-kiong, or hayfluen, known to us by the name of Hyfon Tea, fo called after an Eaft-India merchant of that name, who firft imported it into Europe. The leaves are clofely curled and fmall, of a green colour, verging towards blue '.

' Du Halde's Hiftory of China, Vol. IV. p. 21. Ofbeck's Voyage to China, Vol. I p. 246, et feq.

² The Chinefe have another kind of Hyfon Tea, which they call Hyfon-utchin, with narrow fhort leaves. Another fort of green Tea they name Go-bé, the leaves of which are narrow and long.

111. Singlo,

III. Singlo, or fonglo, which name it receives, like many other Teas, from the place where it is cultivated.

II. The bohea Teas.

1. Soochuen, or futchong, by the Chinefe called faatyang, and faĉt-chaon, or fu-tyann, is a fuperior kind of long-fou Tea. It imparts a yellowifh green colour, by infufion [1].

11. Camho, or foumlo, called after the name of the place where it is gathered; a fragrant Tea with a violet fmell. Its infufion is pale.

111. Cong-fou, congo, or bong-fo. This has a larger leaf than the following, and the infufion is a little deeper coloured. It refembles the common bohea in the colour of the leaf [2].

iv. Pckao, pecko, or pekoe, by the Chinefe called back-ho, or pack-ho. It is known by having the appearance of fmall white flowers intermixed with it.

v. Common bohea, called moji by the Chinefe, confifts of leaves of one colour [3].

[1] Padre futchong has a finer tafte and fmell than the common futchong. The leaves are large and yellowifh, not rolled up, but expanded, and packed up in papers of half a pound each. It is generally conveyed by caravans into Ruffia. Without much care, it will be injured at fea. This Tea is rarely to be met with in England.

[2] There is a fort of Tea called lin-kifam, with narrow rough leaves. It is feldom ufed alone, but mixed with other kinds. By adding it to congo, the Chinefe fometimes make a kind of pekoe Tea. Ofbeck's Voyage to China, Vol. I. p. 249.

[3] The beft bohea Tea is named by the Chinefe tao-kyonn. An inferior kind is called An-kai, from a place of that name. In the diftriĉt of Honam near Canton, the Tea is very coarfe, the leaves yellow or brownifh, and the tafte the leaft agreeable of any. By the Chinefe it is named Honam té, or Kuli té.

III. There

III. There has alſo been imported a ſort of Tea, in balls, of a different form from any of the preceding, made up into cakes or balls of different ſizes, by the Chineſe called Poncul-tcha.

I. The largeſt kind of this cake Tea, that I have ſeen, weighs about two ounces ; the infuſion and taſte reſemble thoſe of good bohea Tea. .

II. Another ſort, which is a kind of green Tea, is called tio tè : it is rolled up in a round ſhape, about the ſize of peas, and ſometimes as large as a nutmeg.

III. The ſmalleſt kind done in this form is called gun-powder Tea.

IV. Sometimes the ſucculent Tea leaves are twiſted into cords like packthread, about an inch and a half or two inches long ; and uſually three of theſe are tied together at the ends by different-coloured ſilk threads. Theſe reſemble little bavins, one of which might ſuffice for tea for one perſon. I have ſeen them both of green and bohea Tea.

The Chineſe likewiſe prepare an extract from Tea, which they exhibit as a medicine diſſolved in a large quantity of water, and aſcribe to it many powerful effects in fevers and other diſorders, when they wiſh to procure a plentiful ſweat. This extract is ſometimes formed into ſmall cakes, not much broader than a ſixpence, ſometimes into rolls of a conſiderable ſize.

That there is only one ſpecies of Tea tree, has already been mentioned (Sect. I.) from which all the varieties of Tea are procured.

BOHEA TEA.

procured. Kæmpfer, who is of this opinion, attributes the difference of Teas to the foil and culture of the plant, age of the leaves when gathered, and method of curing them'. Thefe circumftances will feverally have more or lefs influence; though whether they account for all the varieties obfervable in Tea may be doubted. The bohea Tea trees, now introduced into many botanic gardens near London, exhibit very obvious varieties. The leaves are of a deeper green colour, and not fo deeply ferrated; the ftalk is ufually of a darker colour, and the whole fhrub appears lefs luxuriant than that reprefented in £ the annexed plate of the bohea Tea; but the botanical cha-racters are the fame.

I infufed all the forts of green and bohea Teas I could pro-cure, and expanded the different leaves on paper, to compare their refpective fize and texture, intending thereby to difcover their age. I found the leaves of green Tea as large as thofe of bohea, and nearly as fibrous; which would lead one to fufpect, that the difference does not fo much depend upon the age, as upon the other circumftances.

We know that in Europe the foil, culture, and expofure, have great influence on all kinds of vegetables: but the fame fpecies of plants differ in the fame province, and even in the fame diftrict; and in Japan, and particularly along the continent of China, it muft be much more confiderable, where the air is in fome parts very cold, in others moderate, or warm almoft to an extreme. I am perfuaded that the method of preparation

' This renders what has been obferved at the conclufion of SECT. I. more probable.

muft

muſt alſo have no little influence. I have dried the leaves of
ſome European plants in the manner deſcribed (Sect. VI.)
which ſo much reſembled the foreign Tea, that the infuſion
made from them has been ſeen and drunk without ſuſpicion.
In theſe preparations which I made, ſome of the leaves retained a perfect curl, and a fine verdure like the beſt green
Tea; and others cured at the ſame time were more like the
bohea [1].

I would not, however, lay too much ſtreſs upon the reſult
of a few trials, nor endeavour to preclude further enquiries
about a ſubject which at ſome future period may prove of
more immediate concern to this nation.

We might ſtill try to diſcover whether other arts, than are yet
known here, are not uſed with Tea before its exportation from
China, to produce the difference of colour [2], and flavour [3],
peculiar to different ſorts. An intelligent friend of mine informs
me, that in a ſet of Chineſe drawings, in his poſſeſſion, repre-
ſenting the whole proceſs of manufacturing Tea, there are in one
ſheet the figures of ſeveral perſons apparently ſeparating the

[1] A certain moderate degree of heat preferred the verdure and flavour better
than a haſty exſiccation. In the firſt caſe, it is neceſſary to repeat the roaſting
oftener.

[2] Infuſions of fine bohea Teas do not differ a great deal in colour from thoſe of
green. To ſpirit they equally impart a fine deep green colour.

[3] I am informed by intelligent perſons, who have reſided ſome time at Canton,
that the Tea about that city affords very little ſmell whilſt growing. The ſame is
obſerved of the Tea plants in England; and alſo of the dried ſpecimens from China.
We are not hence to conclude, that art alone conveys to Teas when cured the ſmell
peculiar to each kind; for our vegetables, graſſes for inſtance, have little or no
ſmell till dried, and made into hay.

different

OLEA FRAGRANS.

J. Miller del. et fc:

different kinds of Tea, and drying it in the fun, with feveral
bafkets ftanding near them filled with a very white fubftance,
and in confiderable quantity. To what ufe this may be applied
is uncertain, as well as what the fubftance is; yet there is no
doubt, he thinks, that it is ufed in the manufacturing of Tea,
as the Chinefe feldom bring any thing into their pieces but fuch
as relate in fome refpect to the bufinefs before them.

We are better acquainted with a vegetable fubftance which
has been employed by the Afiatics in giving a flavour to Tea.
This is the Olea Fragrans, whofe flowers are frequently to be
met with in Teas exported from China: and as the plant is now
not unfrequent in the gardens near the metropolis, I am enabled
to give an engraving of the plant and its botanical hiftory [1].

OLEA FRAGRANS.—Sweet-fcented Olive.

Clafs and Order.

DIANDRIA MONOGYNIA.

Generic Character.

Cor. 4. fida : laciniis fubova-tis. *Drupa* monofperma.	*Corol.* 4. cleft : fegments fomewhat ovate. *Drupe*, one-feeded.

[1] See Thunberg's Flora Japonica, from which work the Botanic Hiftory of the
Olea Fragrans is chiefly taken.

Specific

Specific CharaEler and Defcription
from THUNBERG.

OLEA *fragrans* foliis lanceo-
latis ferratis, pedunculis latera-
libus aggregatis unifloris. *Thunb.*
Fl. Japon. p. 18, t. 2.
Caulis, arboreus vaftus.
Rami et ramuli trichotomi,
obfolete tetragoni, glabri.

Folia decuffata, petiolata ob-
longa, acuta, ferrata, margini
fubreflexo, parallelo-nervofa, re-
ticulata, glabra, fupra faturate
viridia, fubtus pallidiora, paten-
tia, in ramulis frequentia, di-
gitalia.

Petioli femiteretes, fulcati,
glabri, femiunguiculares.

Flores in ramulis umbellato-
aggregati, circiter 6 vel 8, pe-
dunculati.

OLIVE *fweet-fcented* with
lanceolate ferrated leaves, peduncles lateral, cluftered, one-
flowered.
Stem, a vaft tree.
Branches both large and fmall
trichotomous, faintly four-cor-
nered and fmooth.

Leaves growing crofs-wife on
leaf-ftalks, oblong, acute, ferrated, edge fomewhat turned
back, ribs parallel, reticulated,
fmooth above, of a deep green
colour, paler beneath, fpreading on the fmall branches, numerous, about the length of
the finger.

Leaf-ftalks, flat on one fide,
round on the other, grooved,
fmooth, half the length of the
finger nail.

Flower on the fmall branches
in cluftered umbels, about 6 or
8 together, ftanding on peduncles.

Pedunculi

Pedunculi filiformes, uniflori, glabri, albidi, unguiculares.

Perianthium, 1-Phyllum, minimum, obfolete 4-dentatum, albidum, glabrum.

Corolla, 1-petala, rotata, flavo-alba; *Tubus* fubnullus; *Limbus* patens, quadrifidus: laciniæ ovatæ, obtufæ, concavæ, craffiufculæ.

Filamenta duo, ori tubi inferta, alba, breviffima.

Antheræ ovatæ, grandiufculæ, didymæ, flavefcentes.

Germen fuperum, oblongum, glabrum.

Stylus filiformis.

Stigmata fimplicia, acuta.

Flower-ftalks filiform, one-flowered, fmooth, whitifh, a finger nail in length.

Perianthium, one-leaved, very minute, faintly four-toothed, whitifh and fmooth.

Corolla of one petal, wheel-fhaped, of a yellowifh-white colour; *Tube* fcarce any; *Limb* fpreading, quadrifid, fegments ovate, obtufe, concave, thickifh.

Filaments two, inferted into the mouth of the tube, white, very fhort.

Antheræ ovate, fomewhat large, double, yellowifh.

Germen above, oblong, and fmooth.

Style filiform.

Stigmata, fimple and pointed.

Sir George Staunton, in his Embaffy to China, Vol. II. p. 467, defcribes another Plant, whofe flowers are ufed for the purpofe of fcenting Tea. The flower refembles the dog-rofe, and the leaves thofe of Tea; hence the Chinefe call it Chawhaw, or Flower of Tea. A Plate of this Plant is annexed, with the following defcription, which this very accurate and learned

learned traveller has obligingly permitted me to introduce
here.

" A Plant very like the Tea flourished at this time on the sides
and the very tops of mountains, where the soil consisted of
little more than fragments of stone crumbled into a sort of
coarse earth by the joint action of the sun and rain. The
Chinese call this plant Cha-whaw, or Flower of Tea, on account of the resemblance of one to the other; and because its
petals, as well as the entire flowers of the Arabian jessamine,
are sometimes mixed among the Teas, in order to increase
their fragrance.

" This plant, the Cha-whaw, is the Camellia Sesanqua of the
botanists, and yields a nut, from whence is expressed an escu-
lent oil equal to the best which comes from Florence. It is
cultivated on this account in vast abundance; and is particularly valuable, from the facility of its culture, in situations fit
for little else." It is delineated on the opposite page.

As green Tea is by some suspected to have been cured on
copper, they have attributed the verdure to be derived from
that metal (Sect. VII.); but, if there were any foundation for this supposition, the volatile alkali, mixed with an
infusion of such Tea, would detect the least portion of copper,
by turning the infusion blue [1].

[1] The hundredth part of a grain of copper, dissolved in a pint of liquor, strikes
a sensible blue with volatile alkalies. Neumann's Chemistry, by Lewis, p. 62.
The finest imperial and bloom Teas shewed no sign of the presence of this metal by
experiment.

Others

Camellia Sesanqua L.

Others have, with lefs propriety, attributed the verdure to green copperas [1]; but this ingredient, which is only falt of iron, would immediately turn the leaves black, and the infufion made from the Tea would be of a deep purple colour [2].

Is it not more probable, that fome green dye, prepared from vegetable fubftances, is ufed for the colouring?

[1] See Short on Tea, p. 16. Boerhaave attributed the verdure of green Tea to this fubftance.

[2] "It is confidently faid in the country, that no plates of copper are ever employed for that purpofe. Indeed, fcarcely any utenfil ufed in China is of that metal, the chief application of which is for coin. The earthen or iron plates are placed over a charcoal fire, which draws all remaining moifture from the leaves, rendering them dry and crifp." Sir G. Staunton's Embaffy, Vol. II. p. 465.

SECTION

SECTION IX.

DRINKING OF TEA.

Neither the Chinese, nor natives of Japan, ever use Tea before it has been kept at least a year; because when fresh it is said to prove narcotic, and to disorder the senses [1]. The former pour hot water on the Tea, and draw off the infusion in the same manner as is now practised in Europe; but they drink it simply without the addition of sugar or milk [2]. The Japanese reduce the Tea into a fine powder, by grinding the leaves in a hand-mill; they then mix them with hot water into a thin pulp, in which form it is sipped [3], particularly by the nobility and rich people. It is made and served up to company in the following manner : the Tea-table furniture, with the powdered Tea inclosed in a box, are set before the company, and the cups are then filled with hot water, and as much of the powder as might lie on the point of a moderate-sized knife is taken out of the box, put into each cup, and then stirred and mixed together with a curious denticulated instru-

[1] Kæmpfer, Amœnit. Exot. p. 625. History of Japan, Vol. II. App. p. 10. 16.
[2] Ofbeck's Voyage to China, Vol. I. p. 299.
[3] This is called koitsjaa, that is, thick Tea, to distinguish it from that made by infusion.

ment

ment till the liquor foams, in which ftate it is prefented to the company, and fipped while warm [1]. From what Du Halde relates, this method is not peculiar to the Japanefe, but is alfo ufed in fome provinces of China [2].

The common people, who have a coarfer Tea (SECT. VI. III.) boil it for fome time in water, and make ufe of the liquor for common drink. Early in the morning the kettle, filled with water, is regularly hung over the fire for this purpofe, and the Tea is either put into the kettle inclofed in a bag, or, by means of a bafket of a proper fize, preffed to the bottom of the veffel, that there may not be any hindrance in drawing off the water. The Bantsjaa Tea (SECT. VI. III.) only is ufed in this manner, whofe virtues, being more fixed, would not be fo fully extracted by infufion.

And indeed Tea is the common beverage of all the labouring people in China : one fcarcely ever fees them reprefented at work of any kind, but the Tea pot and Tea cup appear as their accompaniments. Reapers, threfhers, and all who work out of doors, as well as within, have thefe attendants [3].

To make Tea, and to ferve it in a genteel and graceful manner, is an accomplifhment, in which people of both fexes in Japan are inftructed by mafters, in the fame manner as Europeans are in dancing, and other branches of polite education.

---

[1] An inferior kind of Tea is infufed, and drank in the Chinefe manner. SECT. VI. II. and SECT. IX. I.

[2] Hiftory of China, Vol. IV. p. 22.

[3] In public roads, and in all places of much refort in Japan, and even in the midft of fields and frequented woods, Tea booths are erected; as moft travellers drink fcarcely any thing elfe upon the road. Kæmpfer's Hiftory of Japan, by Scheuchzer, Fol. Vol. II. p. 428.

II   ·   SECTION

SECTION X.

SUCCEDANEA.

CURIOSITY and intereſt would mutually induce the Europeans to make the moſt diligent enquiries in order to diſcover the real Tea ſhrub, or a ſubſtitute in ſome other vegetable moſt reſembling it, Simon Paulli, a celebrated phyſician and botaniſt at Copenhagen, was the firſt who pretended to have diſcovered the real Tea plant in Europe. By opening ſome Tea leaves, he found them ſo much like thoſe of the Dutch myrtle[1], (Flor. Su. 907.) that he obſtinately maintained they were productions of the ſame ſpecies of Tea ; though he was afterwards refuted by ſeveral botaniſts in Europe, and by the ſpecimens ſent to him, and to Dr. Mentzel of Berlin, from the Eaſt-Indies, by Dr. Cleyer[2].

---

[1] Myrica Gale. Goule, Sweet Willow, or Dutch Myrtle. Hudſon's Fl. Angl. p. 368. Linn. Syſtem. Natur. Vol. III. p. 651. A plant of peculiar fragrance, found in the North of England, Brabant, and other Northern countries. Simon de Molingriis was the firſt who oppoſed this opinion of Simon Paulli, by ſhewing the difference betwixt this ſpecies of myrtle and the oriental Tea. See alſo Wilh. Leyl. epiſt. apud Sim. Paulli comment. &c.

[2] Figures of the ſame were publiſhed in the Acta Haffnienſia, and German Ephemerides, Dec. 11. Ann. iv.

Father

Father Labat next thought he had difcovered the real Tea-plant in Martinico [1], agreeing, he fays, in all refpects with the China fort. He pretends alfo to have procured Tea feeds from the Eaft Indies, and to have raifed the plant in America ; but, from his own account, this fuppofed Tea appears to be only a fpecies of Lyfimachia, or what is called Weft-India Tea [2].

Many other pretended difcoveries of the Oriental Tea-tree have been related ; all which have proved erroneous, when properly enquired into. The genus of plant, called by Kæmpfer Tfubakki [3], has the neareft refemblance to it. The leaves of feveral European herbs have been ufed at different times as fubftitutes for Tea, either from fome fimilarity in the fhape of the leaves, or in the tafte and flavour ; among thefe, two or three fpecies of

---

[1] Nouveau Voyage aux Iles d l'Amerique, Paris, 1721, 12mo. 6 vol.

[2] This fhrub I have frequently met with in the Weft-Indies.

[3] Two fpecimens of this plant are now in the phyfic garden at Upfal. About the year 1755, they were brought over from China by M. Lagerftrom, a director of the Swedifh Eaft-India Company, under the fuppofition of being Tea-plants, till they appeared in bloffom, when they proved to be this fpecies of Tfubakki, called by Linnæus, Camellia. Spec. Plant. p. 982. This celebrated Naturalift fays, "That the leaves of his Camellia are fo like the true Tea, that they would deceive the moft fkilful botanift ; the only difference is, that they are a little broader. Amœnit. Academ. Vol. VII. p. 251. See alfo Ellis's Directions for bringing over foreign Plants, p. 28. A Camellia was brought in 1771 from China in good health ; the leaves of this fhrub end in a double obtufe point (obtufely emarginated) like thofe of the Tea tree, which makes them ftill more liable to be miftaken for thofe of the latter. Kæmpfer obferves, that the leaves of a fpecies of Tfubakki are preferved, and mixed with Tea, to give it a fine flavour. Amœnit. Exotic. p. 358. It is now a common plant in the green-houfes about London.

H 2 Veronica

Veronica are particularly recommended [1], befides the leaves of fage [2], myrtle [3], betony [4], floe [5], agrimony, wild rofe [6], and many others [7]. Whether any of thefe are really more falutary

or

[1] Mich. Frid. Lochner, de novis Theæ et Coffeæ Succedaneis. Hall. 1717. 4to. Veronica officinalis. Flor. Suec. p. 12. Veronica Chamædr. Fl. Suec. p. 18. Pechlin Theophilus bibaculus, Franckfort. 1684. Francus, de Veronica vel Theezantem. Coburg. 1690. 12mo. 1700. 12mo. Paris, fub titulo, le Thè de l'Europe. 1704 and 1707, 12mo. Frid. Hoffman de infufi Veronicæ efficacia præferenda herbæ Theæ, Hall. an. 1694. 4to.

[2] Fr. Afforty & Jof. de Tournefort ergo potus ex Salvia falubris, 1695. Wedel, de Salvia, 4to. 1707. Jena. Paulini nobilis falvia Ang. Vindel. an. 1688. 8vo.

[3] Simon Paulli de abufu Theæ et Tabaci. Strafburg, 1665. Lond. 1746.

[4] Botanical writers celebrate this herb for its many virtues; hence arofe the Italian proverb, " *Vende la tonica, et compra la Betonica.*"

[5] In the year 1776, an act was paffed for the more effectual prevention of the manufacturing of afh, elder, floe, and other leaves, in imitation of Tea; and to prevent frauds in the revenue of Excife in refpect to Tea, 17 George III. chap. 29, being an amendment of the act 4 George II. intituled, " An Act to prevent Frauds in the Revenue of Excife with refpect to Starch, Coffee, Tea, and Chocolate. In the Appendix, from Sir George Staunton's Embaffy to China, this is particularly detailed.

[6] Jofeph Serer Lettera fopra la bevanda del Caffé Europæo, Veron. An. 1730. Rofe leaves are here fubftituted for thofe of Tea. Godofred. Thomafius Thea ex Rofis in Cent III. Nat. curiofor. n. 199. See alfo Cent. vij. obf. 15. by J. A. Fifcher.

[7] See Neumann's Chemiftry, by Lewis, p. 375. J. Adrian. Slevogt, De Thea Romana et Silefiaca, an. 1721. Aignan. le prêtre Medecin, avec un Traité du Caffé, et du Thé de France. Paris. an. 1696. 12mo. This author, whofe name is probably corrupted, prefers balm leaves to thofe of the Afiatic Tea.

M. Fr. Lockner, de novis et exoticis Thee et Cafe fuccedaneis Noriberg. 1717. 4to. Et in Eph. Nat. Cur. Cent. vj.

J. Franc. Nic. Faber, de Thea Helvetica, Bafil. 1715. 4to.

J. Georg. Siegefbeck, de Theæ et Caffeæ fuccedaneis in Kanoldiana collectione, an. 1722. Jan.

Zanichelli

or not, is undetermined; and we now find, that from the palace to the cottage every other fubftitute has yielded to the genuine Afiatic Tea [1].

Zanichelli obzervazioni intorno all abufodella Coffea ed alla vertute di innuovo Te-Venegiano. Venez. 1755. 4to.
K. Collegii medici Rundgiorelfe om den mifbruk fom Thee, och Caffe drickande är unders kaftot, famt anwifning pa Swenka örter, at Brucka i ftälle for Thee Stokholm, 1746. 4to.
Conf. Murray, appar. Medicam. Vol. IV. p. 232. & feq.
[1] In fome parts of Europe, however, Tea is yet a ftranger. See Brydone's Tour through Sicily and Malta, Let. 6.

SECTION

SECTION XI.

PRESERVING THE SEEDS FOR VEGETATION.

Many attempts to introduce the Tea-tree into Europe have proved unfuccefsful, owing to the bad ftate of the feeds when firft procured, or to a want of judgement in preferving them long enough in a ftate capable of vegetation. If this complaint arife from the firft caufe, future precautions about fuch feeds will be in vain; it is therefore neceffary to procure frefh, found, ripe feeds, white, plump, and moift internally.

Two methods of preferving the feeds have put us in pof-feffion of a few young plants of the true Tea-tree of China; one is, by inclofing the feeds in bees wax, after they have been well dried in the fun; and the other, by putting them, in-cluded in their pods, or capfules, into very clofe cannifters made of tin and tutenague[1].   But

[1] See Directions for bringing over feeds and plants from the Eaft-Indies, by J. Ellis, F. R. S. &c. in which particular directions are given, both to choofe the proper feeds, and to preferve them in the beft manner for vegetation. See alfo the Naturalift's and traveller's companion, containing inftructions for difcovering and preferving objects of natural hiftory, Sect. III. We may obferve here, that the beft method of bringing over the parts of flowers intire is to put them in bottles of fpirit of wine, good rum, firft runnings, or brandy. In the directions, &c. above-mentioned the learned naturalift has not recommended this eafy method of preferving the parts of fructification; but in a future edition, I am informed he purpofes to do it. Flowers of the Illicium Floridanum, or ftarry annifeed tree, publifhed in the laft volume of Philofophical Tranfactions (LX.) were fent to him in this manner.

In

Boxes for conveying Plants by Sea.

The Box with plants shut down with the openings
at the ends and front left for fresh air.

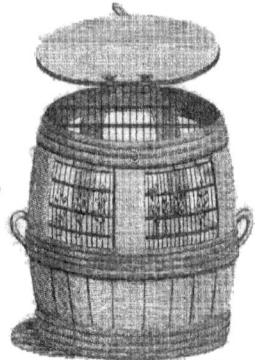

The Cask for sowing seeds with the
openings defended by Wire.

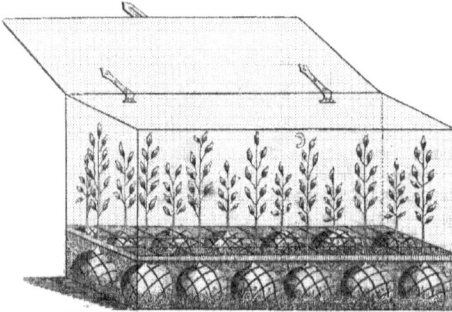

The Inside of the box shewing the manner of securing the roots
of plants surrounded with earth & moss tied with packthread
and fastened crip'd & cryp'd with laths or packthread to keep them
steady.

The Box with divisions for sowing
different seeds in earth & wet moss.

But neither of thefe methods have fuccecded generally, notwithftanding the utmoft care, both in getting frefh feeds, and in fecuring them in the moft effectual manner. The beft method is to fow the ripe feeds in good light earth, in boxes, at leaving Canton; covering them with wire, to prevent rats and other fuch vermin coming to them. The boxes, plans of which are annexed, fhould not be expofed to too much air, nor to the fpray of the fea, if poffible. The earth fhould not be fuffered to grow dry and hard, but a little frefh or rain water may be fprinkled over it now and then; and, when the feedling plants appear, they fhould be kept moift, and out of the burning fun [1]. Moft of the plants now in England were procured by

In a paper by John Sneyd, Efq. inferted in the Tranfactions of the Society for the Encouragement of Arts, Vol. XVI. p. 265, a method of preferving feeds is related, which appears to have been highly fuccefsful; this is merely by packing up feeds in abforbent paper, and furrounding the fame by raifins, or brown moift fugar; which, by experiment, feems to afford that genial moifture requifite to preferve the feeds in a ftate fit for vegetation.

Thouin, in his directions to the unfortunate navigator Pèroufe, recommends the feeds to be placed in alternate layers of earth or fand, in tin boxes, which muft be clofed up exactly, and placed in folid cafes, which fhould be covered with waxed cloth; the boxes fhould be put in a part of the fhip the leaft acceffible to moifture, and the moft fheltered from extreme heat or cold." Vol. I. p. 278.

[1] The carrying of trees cannot be done, with any hope of fuccefs, except in boxes, wherein they may vegetate during the voyage. For this purpofe it is neceffary to have a box forty inches long by twenty broad, and as much in depth, with a dozen holes bored through the bottom, for the fuperabundant water to run off. Its upper part muft be compofed of a triangular frame, upon which lattice work of iron wire muft be fitted, with glazed frames and window fhutters, to keep up a free circulation of air, encreafe the warmth when neceffary, and keep out the cold." Pèroufe's Voyage, Vol. I. p. 283.

thefe

thefe means; and though many of the feedlings will die, yet by
this kind of management we may probably fucceed in bringing
over the moft curious vegetable productions of China, of which
they have an amazing treafure, both in refpect to ufe, fhew, and
variety [1]. If young plants could be procured in China, they
might be fent over in a growing ftate in fome of thefe boxes.

The young Tea-plants in the gardens about London thrive
very well in the green-houfes in winter, and fome bear the
open air in fummer. The leaves of many of them are from
one to three inches long, not without a fine deep verdure;
and the young fhoots are fucculent. It is therefore probable,
that in a few years many layers may be procured from them,
and the number of plants confiderably increafed thereby.

It may not be improper to obferve here, that many exotic
vegetables, like human conftitutions, require a certain period
before they become naturalized to a change of climate; many
plants, which on their firft introduction would not bear our
winters without fhelter, now endure our hardeft frofts; the
beautiful magnolia, among feveral others, is a proof of this

---

[1] Another method has fucceeded with fome North American feeds, by putting
them into a box, not made too clofe, upon alternate layers of mofs, in fuch a manner as to admit the feeds to vegetate, or fhoot their fmall tendrils into the mofs.
In the paffage, the box may be hung up at the roof of the cabin; and, when arrived here, the feeds fhould be put into pots of mold, with a little of the mofs alfo
about them, on which they had lain. This method has procured us feeds in a ftate
fit for vegetation, which had often mifcarried under the preceding precautions;
and therefore might be tried at leaft with Tea and other oriental feeds. In order
to fucceed more certainly, fome of the Tea feeds, in whatever manner they may have
been preferved, fhould be fown when the veffel arrives at St. Helena, and alfo after
paffing the tropic of Cancer, near the latitude of 30 degrees North.

obfervation;

obfervation ; and we have already taken notice (SECT. V.), that the degree of cold at Pekin fometimes exceeds ours. We have hence reafon to expect, that the Tea-tree may in a few years be capable of bearing our climate, or at leaft that of our colonies ; at length thrive, as if indigenous to the foil ; and, were labour cheaper, become an article of export ', like the common potatoe, for which we are indebted to America, or Spain ². It is, however, better fuited for the climates of

the

' The high price of labour in this country may prove the principal objection to this profpect. In China provifions are very cheap. Ofbeck fays, that a work-man who lives upon plucking of Tea-leaves, will fcarce be able to get more than one penny a day, which is fufficient to maintain him. Voyage to China, Vol. I. p. 298.

² The following extract from Gerard's Herbal, p. 780. ed. 1636. though foreign to the fubject of this Effay, is fo curious, that it may not be deemed im-proper to tranfcribe it. " Potatoes grow in India, Barbarie, Spaine, and other hot regions, of which I planted diuers rootes (that I bought in the Exchange in London) in my garden, where they flourifhed untill winter, at which time they perifhed and rotted." At this date, he adds, " they were roafted in the afhes ; fome, when they be fo roafted, infufe them, and fop them in wine ; and others, to give them the greater grace in eating, do boile them with prunes, and fo eate them. And likewife others dreffe them (being firft roafted) with oile, vinegar, and falt, every man according to his own tafte and liking."

" Thefe rootes (he obferves) may ferue as a ground or foundation wheron the cunning confectioner, or fugar-baker, may worke and frame many comfortable delicate conferves, and reftorative fweete meates."

In 1664 J. Fofter publifhed his " England's Happynefs increafed by a Plantation of Potatoes," 4to.

" Captain Hawkins is faid to have brought this root from Santa Fè, in New Spain, A. D. 1565. Sir Walter Rawleigh foon after planted it on his lands in Ireland ; but, on eating the apple, that it produced, which is naufeous and unwholefome, he had nearly configned the whole crop to deftruction. Luckily the fpade difcovered the real potatoe, and the root became rapidly a favourite eatable. It continued, how-ever, for a long time to be thought rather a fpecies of dainty than of provifion ;

I   nor,

the Southern parts of Europe, and America; but hitherto
it has not been cultivated in an extensive manner, in either of
these quarters of the world ; nor is it likely ever to be, whilst it
can be procured from Asia at the present reduced price. It
was introduced into Georgia about the year 1770. Hence the
ingenious author of Ouabi (Mrs. Morton) in her recent poem of
Beave-hill, in describing the products of this province, introduces the exotic of China :

" Yet round these shores prolific plenty twines,
" Stores the thick field, and swells the clustering vines ;
" A thousand groves their glossy leaves unfold,
" Where the rich orange rolls its ruddy gold,
" *China's green shrub,* divine Magnolia's bloom,
" With mingling odours fling their high perfume."

It is indeed probable that the North American summers, in
the same latitude with Pekin, would suit this Tree better than
ours ; for, in China and some parts of North America, the
heat in summer is such, that vegetables make quicker and more
early shoots, whereby they have time to acquire sufficient
strength and firmness before the winter commences : but, in
England, the tender shoots are pushed forth late, and, winter soon
after succeeding, they often perish, in a degree of cold much
less severe than at Pekin, or in colder latitudes of North
America.

nor, till the close of the 18th century, was it supposed capable of guarding the
country where it was fostered, from the attacks of famine." Andrews's History,
Vol. I. p. 408. Comp. Mocquet's Travels, p. 54.

Shakespeare, very early also in this century, mentions this root in the Merry
Wives of Windsor, one edition of which, in 4to. was printed in 1619. Vide Scene
III. Falstaff.

THE

THE

MEDICAL HISTORY

OF

TEA.

PART II.

SECTION I.

As the cuftom of drinking Tea is become general, every perfon may be confidered as a judge of its effects, at leaft fo far as it concerns his own health ; but, as the conftitutions of mankind are various, the effects of this infufion muft be different alfo, which is the reafon that fo many opinions have prevailed upon the fubject.

Many, who have once conceived a prejudice againft it, fuffer it to influence their judgement too far, and condemn the cuftom as univerfally pernicious. Others, who are no lefs biaffed

I 2

on

on the other extreme, would make their own private experience a ſtandard for that of all, and aſcribe the moſt extenſive virtues to this infuſion. This contrariety of opinion has been particularly maintained among phyſicians [1]; which will ever be the caſe, while mere ſuppoſitions are placed in the room of experiments and facts impartially related.

Some phyſicians, however, avoid both extremes; who, without commending it, or decrying it univerſally, admit its uſe, while they are not inſenſible of the injuries it may produce. It requires no ſmall ſhare of ſagacity to fix the limits of good and harm in the preſent caſe: multitudes of all ages, conſtitutions, and complexions, drink it freely, during a long life, without perceiving any ill effects. Others, again, ſoon experience many inconveniences from drinking any conſiderable quantity of this infuſion.

It is difficult to draw certain concluſions from experiments made on this herb. The parts which ſeem to produce theſe oppoſite effects are very fugitive. We become acquainted chiefly with the groſſer parts by analyſis. I made the following experiments with conſiderable care; but, I own, they inform us not ſufficiently wherein conſiſts that grateful relaxing ſedative property, which proves to the generality of mankind ſo refreſhing; nor from whence it is, that others feel from the pleaſing beverage ſo many diſagreeable effects. Accurate obſervation would inſtruct us in this difficult inveſtigation, more than ſimple experiments on the ſubject itſelf.

[1] Compare Joh. Ludov. Hannemane de potu calido in Miſcell. curioſ. Simon Paulli de abuſu Theæ et Tabaci. Tiſſot on the diſeaſes of literary and ſedentary perſons, &c. with Waldſmick, in Diſput. var. argum. &c.

EXPE-

EXPERIMENT I.

I took an equal quantity of an infufion of fuperfine green Tea, and of common bohea Tea, made equally ftrong; and alfo the fame quantity of the liquor remaining after diftillation (Sect. III. 1.), and of fimple water; into each of which, contained in feparate veffels, I put two drachms of beef, that had been killed about two days.

The beef, which was immerfed in the fimple water, became putrid in forty-eight hours; but the pieces in the two infufions of Tea, and in the liquor remaining after diftillation, fhewed no figns of putrefaction, till after about feventy hours [1].

EXPERIMENT II.

Into ftrong infufions of every kind of green and bohea Tea that I could procure, I put equal quantities of falt of iron (fal martis), which immediately changed the feveral infufions into a deep purple colour [2].

It

[1] See Percival's Experimental Effays, p. 119, et feq. wherein many ingenious experiments and obfervations are related.

[2] In this experiment, four ounces of infufion were drawn from two drachms of each kind of Tea, and one grain of fal martis added to the refpective infufions.

See

It is evident from thefe experiments, that both green and bohea Tea poffefs an antifeptic (Experiment I.), and aftringent power (Experiment II.), applied to the dead animal fibre.

See Neumann's Chemiftry by Lewis, page 377. Short, on the Nature and Properties of Tea, p. 29. The firft author I have met with, that tried this experiment, was J. And. Hahn, who wrote in the year 1722. De herbæ exoticæ Theæ infufo, ejufque ufu et abufu, Erford, 4to. Though it fhould be premifed, that Nic. de Blegny, who publifhed his work, intituled, *Le bon ufage du Thé*, &c. in 1680, takes notice of the aftringency of Tea, from which quality he deduces many of its virtues. Vid. Act. Eruditor. V. vi. page 49. Ann. 1688.

SECTION

SECTION II.

NEVERTHELESS, as I have often obferved that drinking Tea, particularly the moft highly-flavoured fine green, proves remarkably relaxing to many perfons of tender and delicate conftitutions, I was induced to profecute my enquiries farther.

1. To this end I diftilled half a pound of the beft and moft fragrant green Tea with fimple water [1], and drew off an ounce of very odorous and pellucid water, free from oil, and which on trial (SECTION I. EXPERIMENT II.) fhewed no figns of aftringency.

2. That part of the liquor which remained after diftillation, was evaporated to the confiftence of an extract; it was flightly odorous, but had a very bitter, ftyptic, or aftringent tafte. The quantity of the extract thus procured weighed about five ounces and a half [2].

EXPERIMENT III.

a. Into the cavity of the abdomen, and cellular membrane of a frog, about three drachms of the diftilled odorous water (No. 1.) were injected.

[1] J. Andr. Hahn takes notice alfo of the odour of the water diftilled from Tea.

[2] The fame author prepared an extract from this Tea, though in a lefs proportion than my experiment afforded, or what Neumann relates from his.

In

In twenty minutes, one hind leg of the frog appeared much affected, and a general lofs of motion and fenfibility fucceeded[1]. The affection of the limb continued for four hours, and the univerfal torpidity remained above nine hours; after this the animal gradually recovered its former vigor.

b. In like manner fome of the liquor remaining after the diftillation of the green Tea (No. 1.) was injected; but this was not productive of any fenfible effect.

EXPERIMENT IV.

a. To the ifchatic nerves laid bare, and to the cavity of the abdomen of a frog, I applied fome of the diftilled odorous water (No. 1. and EXPERIMENT III, 1.). In the fpace of half an hour, the hindermoft extremities became altogether paralytic and infenfible; and in about an hour afterwards the frog died.

b. In like manner I applied the liquor remaining after diftillation (No. 1. and EXPERIMENT III. 2.) to another frog; but no fedative or paralytic effect was obfervable.

[1] These infufum, nervo mufculove ranæ admotum, vires motrices minuit, perdit. Smith, Tentamen inaugurale de actione mufculari. Edinburgh, p. 46. Exper. 36.

3. From

3. From thefe experiments the fedative and relaxing effects of Tea appear greatly to depend upon an odorous fragrant principle, which abounds moft in green Tea, particularly that which is moft highly flavoured [1]. This feems farther confirmed by the practice of the Chinefe, who avoid ufing this plant, till it has been kept at leaft twelve months, as they find when recent it poffeffes a foporiferous and intoxicating quality. (Part I. Sect. IX.)

Thus often under trees fupinely laid,
Whilft men enjoy the pleafure of the fhade,
Whilft thofe their loving branches feem to fpread
To fcreen the fun, they noxious atoms fhed,
From which quick pains arife, and feize the head.
Near Helicon, and round the learned hill
Grow Trees, whofe bloffoms with their odour kill [2].

[1] Two drachms of this odorous water were given to a delicate perfon. He was foon after affected with a naufea, ficknefs, general lownefs, and debility, which continued for fome hours, which he obferves ufually follows the ufe of fuperfine green Tea.

Smelling forcibly at the fame has occafioned fimilar effects upon fome delicate people. Dr Blegny, who wrote in 1680, attributes confiderable virtues to this fragrant odour, which he recommends to be breathed into the lungs, where it acts as a fedative, according to his own relation, producing fleep, and relieving pains of the head. Agreeable to Counfellor De Blegny's experience, I know a lady, frequently troubled with a nervous head-ach, who ufed to hold her head over a hot infufion of Tea, and thus receive the fragrant exhalation, which always affords her the moft inftantaneous and effectual relief.

[2] Arboribus primum certis gravis umbra tributa eft
Ufque adeo, capitis faciant ut fæpe dolores,
Si quis eas fubter jacut profiratus in herbis.
Eft etiam in magnis Heliconis montibus arbos
Floris odore hominem tetro confueta necare.   Lucretius, B. 6.

K   SECTION

SECTION III.

WAVING, however, any attempts to fix with precifion the effects of Tea from thefe experiments alone, let us endeavour to collect from obfervation likewife, fuch facts as may enable us to judge what its effects are on the human frame, and from thence draw the cleareft inferences we can, how far it is falutary or otherwife.

The long and conftant ufe of Tea, as a part of our diet, makes us forget to enquire whether it is poffeffed of any medicinal properties. We fhall endeavour to confider it in both refpects.

The generality of healthy perfons find themfelves not apparently affected by the ufe of Tea: it feems to them a grateful refrefhment, both fitting them for labour and refrefhing them after it. There are inftances of perfons who have drank it from their infancy, to old age; have led, at the fame time, active, if not laborious lives; and yet never felt any ill effects from the conftant ufe of it.

Where this has been the cafe, the fubjects of both fexes were for the moft part healthy, ftrong, active, and temperate. Amongft the lefs hardy and robuft, we find complaints, which are afcribed to Tea, by the parties themfelves. Some complain that after a Tea breakfaft, they find themfelves rather

fluttered;

fluttered; their hands lefs fteady in writing, or any other employment that requires an exact command of fpirits. This probably foon goes off, and they feel no other injury from it.
Others again bear it well in the morning, but from drinking it
in the afternoon, find themfelves very eafily agitated, and
affected with a kind of involuntary trembling.

There are many people who cannot bear to drink a fingle
difh of Tea, without being immediately fick and difordered at
the ftomach : To fome it gives excruciating pain about that
part, attended with general tremours. But in general the moft
tender and delicate conftitutions are moft affected by the free
ufe of Tea; being frequently attacked with pains in the ftomach
and bowels; fpafmodic affections; attended with a copious
difcharge of limpid urine, and great agitation of fpirits on the
leaft noife, hurry, or difturbance.

K 2 SECTION

SECTION IV.

THERE is one circumſtance, however, that renders it more difficult to inveſtigate the certain effects of Tea; which is, the great unwillingneſs that moſt people ſhew, to giving us a genuine account of their uneaſy ſenſations after the free uſe of it; from a conſciouſneſs that it would be extremely imprudent to continue its uſe, after they are convinced from experience that it is injurious.

That it produces watchfulneſs in ſome conſtitutions is moſt certain, when drank at evening in conſiderable quantities. Whether warm water, or any other aqueous liquor, would have the ſame effect, is not certain.

That it enlivens, refreſhes, exhilarates, is likewiſe well known. From all which circumſtances it would ſeem, that Tea contains an active penetrating principle, ſpeedily exciting the action of the nerves; in very irritable conſtitutions, to ſuch a degree as to give very uneaſy ſenſations, and bring on ſpaſmodic affections: in leſs irritable conſtitutions, it rather gives pleaſure, and immediate ſatisfaction, though not without occaſionally producing ſome tendency to diſagreeable tremours and agitation.

The

The finer the Tea, the more obvious are these effects. It is perhaps for this, amongst other reasons, that the lower classes of people, who can only procure the most common, are in general the least sufferers. I say, in general, because even amongst them there are many who actually suffer much by it; they drink it as long as it yields any taste, and, to add to its flavour, for the most part hot; and thus the quantity which they take, and the degree of heat in which it is drank, conspire to produce in them, what the finer kinds of Tea effect in their superiors.

It ought not, however, to pass unobserved, that in a multitude of cases the infusions of our own herbs, sage for instance, mint, baum, even rosemary, and valerian itself, will sometimes produce similar effects, and leave that sensation of emptiness, agitation of spirits, flatulence, spasmodic pains, and other symptoms, that are met with in people, the most of all others devoted to Tea.

Besides the injuries which the stomach sustains, by taking the infusion of Tea extremely hot; it is not improbable but the teeth also are affected by it. Professor Kalm, in his Travels into North America, observes, that such of the inhabitants as took their Tea and food in general, in this state, were frequently liable to lose half their teeth at the age of twenty, without any hopes of getting new ones. This cannot be attributed to the variations of weather in that clime, because the Indians who enjoy the same air, but take their viands almost cold, were to a great age possessed of fine white teeth;

as

as were likewife the Europeans who firft fettled in America, before the ufe of Tea became general. It was no lefs remarkable, that the Indian women, who had accuftomed themfelves to drink this infufion after the European fafhion, had likewife loft their teeth prematurely, though they had formerly been quite found [1]. Kalm does not appear to fufpect any injury to the teeth from the fugar ufed with the Tea.

[1] Vol. I. p. 282. Ed. 2.

SECTION. V.

MANY, from a fuppofition that Tea was dried in India on copper, have attributed its pernicious properties to this metal; but we have already obferved (Part I. § VIII.), that, if Tea were tinctured with the leaft quantity of copper, it might eafily be detected by chemical experiments.

Some have attributed the injurious qualities of this fafhionable exotic upon the ftomach to the fugar ufually drank with the Tea; but I have had fufficient opportunities of obferving in the Weft Indies the good effects of drinking freely the juice of the fugar-cane, to obviate this objection. I have known feeble emaciated children, afflicted with worms, tumefied abdomen, and a variety of difeafes, foon emerge from their complicated ailments, by drinking large draughts of this fweet liquor, and become healthy and ftrong[1].

" While

[1] In fome parts of Scotland the common people give children large draughts of fugar and water to deftroy worms. See alfo Boerhaav. Elem. Chemiae, Tom. II. p. 160. Hiftorifch Verhaal. &c. inde Voorreeden Bezoar. London, 1715, 8vo. Slare de Sacchar. et lapid. Van. Swieten Commen. v. V. p. 586. Duncan, in his Avis Salutaire, frequently introduces fugar as an agreeable poifon, though he offers no proof in fupport of this epithet. Dr. Robertfon, in his Hiftory of Charles V. Vol. I. p. 401, 8vo. obferves, that " fome plants of the Sugar-cane were brought from Afia; and the firft attempt to cultivate them in Sicily was made about the middle

" While flows the juice mellifluent from the cane,
" Grudge not, my friend, to let thy flaves, each morn,
" But chief the fick and young, at fetting day,
" Themfelves regale with oft-repeated draughts
" Of tepid nectar, and make labour light '."

That there is fomething in the finer green Teas, that produces effects peculiar to itfelf, and not to be equalled by any
other fubftance we know, is, I believe, admitted by all who
have obferved, either what paffes in themfelves, or the accounts
that others give of their feelings, after a plentiful ufe of this
liquor. Nor are the finer kinds of bohea Teas incapable of the
like influence. They affect the nerves, produce tremblings,
and fuch a ftate of body for the time, as fubjects it to be agitated by the moft trifling caufes, fuch as fhutting a door too
haftily, the fudden entrance even of a fervant, and other the
like caufes.

middle of the 12th century. From thence they were tranfplanted into the fouthern
provinces of Spain. From Spain they were carried to the Canary and Madeira Ifles,
and at length into the New World. Ludovico Guicciardini, in enumerating the
goods imported into Antwerp, about the year 1560, mentions the fugar which they
received from Spain and Portugal as a confiderable article of import. He defcribes
that as the product of the Madeira and Canary iflands. Defcritt. de Paefi Baffi,
p. 180, 181. The fugar-cane was either not introduced into the Weft-Indies at that
time, or the cultivation of it was not fo confiderable as to furnifh an article in commerce. In the middle ages, though Sugar was not raifed in fuch quantities, or
employed for fo many purpofes, as to become one of the common neceffaries of life,
it appears to have been a confiderable article in the commerce of the Italian States."
It is, however, well afcertained, that the Sugar Cane is indigenous to South America,
and the Weft-Indies. Mofeley on Sugar, p. 29.

' Granger's Sugar Cane, 4to. p. 109. See alfo p. 9.
" Dulces bibebant ex arundine fuccos.   LUCAN.
Μελι καλαμινον το λεγομενον σακχαρι.   ARRIAN.

I know

I know people of both fexes, who are conftantly feized with great uneafinefs, anxiety, and oppreffion, as often as they take a fingle cup of Tea, who neverthelefs, for the fake of company, drink feveral cups of warm water, mixed with fugar and milk, without the fame inconvenience.

A phyfician, whofe acquaintance I have long been favoured with, and who, with fome others, was prefent when the preceding experiments were made at the college of Edinburgh, has a remarkable delicacy in feeling the effects of a fmall quantity of fine Tea. If drank in the forenoon, it affects his ftomach with an uneafy fenfation, which continues for feveral hours, and entirely takes away his appetite for food at dinner; though at other times, when he takes chocolate for breakfaft, he generally makes a very hearty meal at noon, and enjoys the moft perfect health. If he drink a fingle difh of tea in the afternoon, it affects him in the fame manner, and deprives him of fleep for three or four hours through the fucceeding night; yet he can take a cup of warm water with fugar and milk, without the leaft inconvenience.

It may be remarked that opium has nearly the fame effect upon him as Tea, but in a greater degree; for he informs me, that when he once accidentally took a quantity of the folution of opium, it had not the leaft tendency to induce fleep, but produced a very difagreeable uneafinefs at his ftomach, approaching to naufea. The late celebrated Profeffor Whytt [1], of Edinburgh, affords a ftriking example how injurious the effects of Tea may be upon conftitutions, which I fhall relate in his

[1] Whytt's Works, 4to. p. 642.

own

own words. " I once imagined Tea to be in a great meafure unjuftly accufed; and that it did not hurt the ftomach more than an equal quantity of warm water; but experience has fince taught me the contrary. Strong Tea drunk in any confiderable quantity, in a morning, efpecially if I eat little bread with it, generally makes me fainter before dinner than if I had taken no breakfaft at all; at the fame time it quickens my pulfe, and often affects me with a kind of giddinefs. Thefe bad effects of Tea are moft remarkable when my ftomach is out of order."

SECTION VI.

I am informed likewife by a phyfician, of long and extenfive practice in the city, that he has known feveral inflances of a fpitting of blood having been brought on, by breathing in an air loaded with the fine duft of Tea. It is cuftomary for thofe who deal largely in this article to mix different kinds together, fo as to fuit the different palates of their cuftomers. This is generally performed in the back part of their fhops, feveral chefts perhaps being mixed together at the fame time. Thofe who are much employed in this work are at length very often fufferers by it; fome are feized with fudden bleedings from the lungs or from the noftrils; and others attacked with violent coughs, ending in confumptions.

Thefe circumftances are chiefly brought in fight to prove, that, befides a fedative relaxing power, there exifts in Tea an active penetrating fubftance, which, in many conftitutions, cannot fail of being productive of fingular effects.

An eminent Tea-broker, after having examined in one day upwards of one hundred chefts of Tea, only by fmelling at them forcibly, in order to diftinguifh their refpective qualities, was the next feized with a violent giddinefs, head ach, univer-

L 2 fal

fal fpafms, and lofs of fpeech and memory. By proper affift-ance, the fymptoms abated, but he did not totally recover. For, though his fpeech returned, and his memory in fome degree, yet he continued, with unequal fteps, gradually lofing ftrength, till a partial paralyfis enfued, then a more general one, and at length he died. Whether this was owing to the effluvia of the Tea, may perhaps be doubted. Future accidents may pof-fibly confirm *the fufpicions* to be juft or otherwife.

SECTION

SECTION VII.

A n affiftant to a Tea broker, had frequently for fome weeks complained of pain and giddinefs of his head, after examining and mixing different kinds of Tea : the giddinefs was fometimes fo confiderable, as to render it neceffary for a perfon to attend him, in order to prevent any injury he might fuffer from falling or other accident. He was bled in the arm freely, but without permanent relief; his complaint returned as foon as he was expofed to his ufual employment. At length he was advifed to be electrified, and the fhocks were directed to his head. The next day his pain was diminifhed, but the day after clofed the tragical fcene. I faw him a few hours before he died ; he was infenfible; the ufe of his limbs almoft loft, and he funk very fuddenly into a fatal apoplexy. Whether the effluvia of the Tea, or electricity, was the caufe of this event, is doubtful. In either view the cafe is worthy of attention [1].

A young man of a delicate conftitution, had tried many powerful medicines in vain, for a depreffion of fpirits, which he

---

[1] From thefe inftances of the deleterious effects of Tea, one might be led to fuppofe that the fame unhappy confequences would frequently attend thofe who are employed in examining and mixing different kinds of Tea in China ; but there the Teas are mixed under an open fhed, through which the air has a free current, by which the odour and the duft are diffipated : but in London this bufinefs is ufually done in a back room, confined on every fide.

laboured

laboured under to a degree of melancholy, which rendered his
fituation dangerous to himfelf and thofe about him. I found
he drank Tea very plentifully, and therefore requefted him to
fubftitute another kind of diet; which he complied with, and
afterwards gradually recovered his ufual health. Some weeks
after this, having a large prefent of fine green Tea fent him,
he drank a confiderable quantity of the infufion on that and
the following day. This was fucceeded by his former dejection
and melancholy, with lofs of memory, tremblings, a pronenefs
to great agitation from the moft trifling circumftances, and a
numerous train of nervous ailments. I faw him again, and he
immediately attributed his complaints to the Tea he had drank;
fince which he has carefully denied himfelf the fame indulgence,
and now enjoys his former health.

I have known many other inftances, where lefs degrees of
depreffion, and other complaints depending upon a relaxed irritable habit, have attended delicate people for many years; and
though they have had the advice of fkilful phyficians, yet in
vain have medicines been adminiftered, till the patient has re-
frained from the infufion of this fragrant exotic[1].

[1] Van Swieten, in his Commentaries on Boerhaave's aphorifms, fpeaks of the
effects of Tea and Coffee in the following manner. " Vidi plurimos, his potibus
diu abufos, adeo enervatum corpus habuiffe, ut vix languida membra traherent, ac
plures etiam apoplexia et paralyfi correptos fuiffe." Tom. III. § 1060, p. 362,
de paralyfi.

SECTION

SECTION VIII.

I N treating of this fubftance, I would not be underftood to be either a partial advocate, or a paffionate accufer. I have often regretted that Tea fhould poffefs any pernicious qualities, as the pleafure which arifes from refle&ting how many millions of our fellow-creatures are enjoying at one hour the fame amufing repaft ; the occafions it furnifhes for agreeable converfation ; the innocent parties of both fexes it daily draws together, and entertains without the aid of fpirituous liquors; would afford grateful fenfations to a focial breaft. But juftice demands fomething more. It ftands charged by many able writers, by public opinion, partly derived from experience, with being the caufe of many diforders; all that train of diftempers included under the name of NERVOUS are faid to be, if not the offspring, at leaft highly aggravated by the ufe of Tea. To enumerate all thefe would be to tranfcribe volumes. It is not impoffible but the charges may be partly true. Let us examine them with all poffible candour.

The effe&t of drinking large quantities of any warm aqueous liquor, according to all the experiments we are acquainted with, would be, to enter fpeedily into the courfe of circulation, and pafs off as fpeedily by urine or perfpiration, or the increafe of fome of the fecretions. Its effe&ts on the folid parts of the

conftitution

constitution would be relaxing, and thereby enfeebling. If this warm aqueous fluid were taken in confiderable quantities, its effects would be proportionable; and still greater, if it were fubstituted instead of nutriment [1].

That all infusions of herbs may be confidered in this light feems not unreasonable. The infusion of Tea, nevertheless, has these two particularities. It is not only possessed of a fedative quality (Sect. II. Exp. III. IV.), but also of a confiderable astringency (Sect. II. Exp. II.) ; by which the relaxing power afcribed to a mere aqueous fluid is in fome meafure corrected. It is, on account of the latter, perhaps less injurious than many other infusions of herbs, which, befides a very flight aromatic flavour, have very little if any stypticity, to prevent their relaxing debilitating effects.

Tea, therefore, if not too fine, nor drank too hot, or in too great quantities, is, perhaps, preferable to any other vegetable infusion we know. And if we take into confideration likewife its known enlivening energy, it will appear that our attachment to Tea is not merely from its being costly or fashionable, but from its fuperiority in taste and effects to most other vegetables.

[1] Vide Trattato di Medicina prefervation : Scritto da Carlo Gianella. Veron.. 1751. p. 112. Simon Pauli, who took a pleafure in oppofing the ufe of Tea, indulges himfelf with the irony of the following lines :

Drinσ Wiin and warff,
Drinσ Beer and verdarff,
Drinσ Waater and ftarff :

Or ;

Drink Wine, and profit ;
Drink Beer, and grow thin ;
Drink Water, and die.

SECTION

SECTION IX.

It may be of fome ufe in our inquiries to confider its effects where it has been long and univerfally ufed. Of Japan we know little at prefent: of China we have more recent accounts; from thefe it appears, that Tea of fome kind, coarfer or finer, is drank plentifully by all degrees of people; the general provifion of the lower ranks efpecially is rice, their beverage Tea. The fuperior claffes of people drink Tea; but they likewife partake of animal food, and live freely.

Of their difeafes we know but little, nor what effects Tea may have in this refpect. They feldom or never bleed. The late Dr. Arnot, of Canton, a gentleman who did his profeffion and his country honour, and was in the higheft eftimation with the Chinefe, I am informed, was the firft perfon who could ever prevail upon any of the Chinefe to be blooded [1], be their maladies what they might. It would appear from hence, that inflammatory difeafes were not frequent among them; otherwife a nation, who feem fo fond of life as the Chinefe are reputed to be, would by fome means or other have admitted of this almoft only remedy in fuch cafes. May we infer from hence, that inflammatory difeafes are lefs frequent in China,

---

[1] See Du Halde's hiftory of China, V. III. p. 362. He obferves here, that bleeding is not entirely unknown amongft the Chinefe.

M

than

than in fome other countries, and that one caufe of this may be the conftant and liberal ufe of this infufion ? Perhaps, if we take a view of the ftate of difeafes, as exactly defcribed a century ago, and compare it with what we may obferve at prefent, we may have a collateral fupport for this fuggeftion. If we confider the frequency of inflammatory difeafes in Sydenham's time, who was both a confummate judge of thefe difeafes, and defcribed them faithfully, I believe we fhall find they were then much more frequent than they are prefent ; at leaft, if any deference is due to the obfervations of judicious perfons, who moftly agree, that genuine inflammatory difeafes are much more rare at prefent, than they were at the time when Sydenham wrote. It is true, this difpofition, admitting it be fact, may arife from various caufes ; amongft the reft, it is not improbable, Tea may have its fhare.

SECTION

SECTION X.

Before the ufe of Tea, the general breakfaft in this country confifted of more fubftantial aliment[1]; milk in various fhapes, ale and beer, with toaft, cold meat, and other additions. The like additions, with fack, and the moft generous wines, found their way amongft the higher orders of mankind. And one cannot fuppofe but that fuch a diet, and the ufual exercife they took, would produce a very different ftate of blood and other animal juices, from that which Tea, a little milk or cream, and bread and butter, affords.

It was not the breakfaft only that feems to have contributed its fhare towards introducing a material alteration in the animal fyftem, but the fubfequent regale likewife in the afternoon.

[1] The late Owen Salufbury Brereton, Efq. a gentleman well known among the learned, had in his poffeffion a MS. dated "apud Eltham, menfe Jan. 22, Hen. viij." intituled, "Articles devifed by his Royal Highnefs (the title of Majefty was not given to our Kings till a reign or two after), with Advice of his Council, for the Eftablifh-ment of good Order and Reformation of fundry Errors and Mifufes in his Houfehold and Chambers." In p. 85, "The queen's maids of honour to have a chet loaf, a manchat, a gallon of ale, and a chine of beef, for their breakfafts." Compare the Archæologia, publifhed by the Society of Antiquaries of London, Vol. III. p. 157. Hume's Hiftory of England, Vol. IV. p. 499. Hiftoria delle cofe occorfe nel regno d'Inghilterra in materia del Duca di Notomberlan dopo la morte di Odvardo vi. Venice, 1538.

Tea

Tea is a fecond time brought before company; it is drank by moft people, and often in no very fmall quantities. Before the introduction of this exotic, it was not unufual to entertain afternoon guefts in a very different manner; jellies, tarts, fweetmeats; nay, cold meat, wine, cyder, ftrong ale, and even fpirituous liquors under the title of cordials, were often brought out on thefe occafions, and perhaps taken to excefs, much to the injury of individuals.

This kind of repaft would tend to keep up the natural inflammatory diathefis, which was the refult of vigour, and a plenitude of rich blood, as well as favour difeafes originating from fuch caufes. It feems not unreafonable therefore to fuppofe, that, as the diet of our anceftors was more generous, their exercifes more athletic, and their difeafes more generally the produce of a rich blood, than are obfervable in the prefent times, thefe debilitating effects before-mentioned may in part be attributed to the ufe of Tea, as no caufe appears to be fo general and fo probable.

SECTION

SECTION XI.

I F thefe fuggeftions are admitted, they will affift us in determining when and to whom the ufe of Tea is falutary, and to whom it may be deemed injurious. Thofe, for inftance, who either from a natural propenfity to generate a rich inflammatory blood, or from exercife, or diet, or climate, or all together, are difpofed to be in this fituation: to thefe the ufe of Tea would feem rather beneficial, by relaxing the too rigid folids, and diluting the coagulable lymph of the blood, as a very fenfible and ingenious author very juftly ftyles it [1].

There are idiofyncrafes, certain particularities, which are objections to general rules. There are, for inftance, men of this temperament, ftrong, healthy, vigorous, and with not only the appearance, but the requifites of firm health, to whom a few difhes of Tea would produce the agitations familiar to an hyfteric woman; but this is by no means general: in common they bear it well, it refrefhes them, they endure fatigue after it, as well as after the moft fubftantial viands. Nothing refrefhes them more than Tea, after lafting and vehement exercife. To fuch it is undoubtedly wholefome, and equal at leaft, if not preferable, to any other kind of regale now in ufe.

[1] Philofophical Tranfactions, Vol. LX. 1770. p. 368, & feq.

But,

But, if we confider what may reafonably be fuppofed to happen to thofe who are in the oppofite extreme of health and vigour; that is, the tender, delicate, enfeebled, whofe folids are debilitated, their blood thin and aqueous, the appetite loft or depraved, without exercife, or exercifing improperly; in fhort, where the difpofition of the whole frame is altogether oppofite to the inflammatory; the free and unreftrained ufe of this infufion, and fuch accompanyments, muft unavoidably contribute to fink the remains of vital ftrength ftill lower.

Between thefe two extremes there are many gradations; and, every thing elfe being alike, Tea will in general be found more or lefs beneficial or injurious to individuals, in proportion as their conftitutions approach nearer to thefe oppofite extremes. To defcend into all the particulars would require experience and abilities, more than I can boaft. Suffice it to fay, that, except as a medicine, or after great fatigue, large quantities are feldom beneficial, nor fhould it ever be drank very hot; and, as hath been already mentioned, the finer Tea, the green efpecially, is more to be fufpected than the common or middling kinds.

SECTION

SECTION XII.

THE experiments and obfervations hitherto related render it evident, that Tea poffeffes a fragrant volatile principle, which in general tends to relax and enfeeble the fyftem of delicate perfons, particularly when it is drank hot, and in large quantities. I have known many of this frame of conftitution, who have been perfuaded, on account of their health, to deny themfelves this fafhionable infufion, and received great benefit (SECT. VII.). Others, who have found their health impaired by this indulgence, are unhappily induced to continue it for want of an agreeable fubftitute, efpecially for breakfaft.

But, if fuch cannot wholly omit this favourite regale, they may certainly take it with more fafety, by boiling the Tea a few minutes, in order to diffipate this fragrant principle (SECT. II. 1, and EXP. IV.) which is the moft noxious; and extract the bitter, aftringent, and moft ftomachic part (SECT. II. 2, and EXP. III.) inftead of preparing it in the ufual manner by infufion.

An eminent phyfician in the city, frequently experiencing the prejudicial effects of Tea by drinking it in the ufual form, was induced, from reading a differtation upon this fubject, publifhed fome time fince at Leyden [1], to try the infufion pre-

---

[1] Siftens Obfervationes ad vires Theæ pertinentes. Lugd. Batav. 1769.

pared

pared after another manner. He ordered the Tea to be infufed in hot water, which after a few hours he caufed to be poured off, ftand over night, and to be made warm again in the morning for breakfaft. By this means, he affures me, he can take, without inconvenience, near double the quantity of Tea, which formerly, when prepared in the ufual method, produced many difagreeable nervous complaints.

The fame end is obtained by fubftituting the extract of Tea (SECT. II. 2.) inftead of the leaves. It may be ufed in the form of Tea, by diffolving it in warm water ; and, as the fragrancy of the Tea is in this cafe diffipated, the nervous relaxing effects, which follow the drinking it in the ufual manner, would be in great meafure avoided. This extract has been imported into Europe from China, in flat round dark-coloured cakes, not exceeding a quarter of an ounce each in weight, ten grains of which, diffolved in a fufficient quantity of water, might fuffice one perfon for breakfaft. It might alfo be made here without much expence or trouble (See SECT. II. 2.).

It is remarkable, that in all the forms which Du Halde relates, for adminiftering Tea as a ftomachic medicine among the Chinefe, it is ordered to be boiled for fome time, or prepared in fuch a manner, as to caufe a diffipation of its fragrant perifhable flavour ; which practice, as it feems confonant to experiments here (SECT. II. EXP. III.), may probably have taken its rife in China, from long experience and repeated facts.

SECTION

SECTION XIII.

PERHAPS it will not be deemed foreign to an essay upon this subject, to take a concise view of the manners and dispositions of the Chinese, as we have done of their diseases. Those who are best acquainted with human nature seem to ascribe even to their food, and way of life, as well as to their climate and education, certain propensities at least to vice and virtue; and it may be of use to draw what light we can in these respects, from the character of a people, who have used the infusion of Tea for a long series of years.

They are in general described to be a people of moderate strength of body, not capable of much hard labour, rather feeble when compared with the inhabitants of some nations, excelling in some minute fabricks and manufactures, but exhibiting no proofs of elevated genius in architecture, either civil or military. They are said to be pusillanimous, cunning, extremely libidinous, and remarkable for dissimulation and selfishness [1], effeminate, revengeful, and dishonest [2].

[1] See Anson's Voyage round the World, 8vo. p. 366, and many later authorities.

[2] See likewise Du Halde's History of China, Vol. II. p. 75, 130, et seq. Les Lettres Curieuses et Edifiantes des Jesuites.

N It

It would be unjuſt to aſcribe all theſe qualities to their manner of living: other cauſes have undoubtedly their ſhare:
but it may be ſuſpected, that the manner of life, or kind of
diet, that tends to debilitate, virtually contributes to the increaſe of the meaner qualities. When force of body is want-
ing, cunning often ſupplies its place; and if not regulated
by other principles, it would diſcover its effects more univerſally; and thus will take place whether the debility is natural,
or acquired by a diet that enfeebles the body. That there is a
probity, fortitude, and generoſity, in female minds, not inferior
to the like qualities poſſeſſed by the other ſex, is moſt certain;
but that it is generally ſo may perhaps be doubted;

> though both
> Not equal, as their ſex not equal ſeem'd;
> For contemplation he and valour form'd,
> For ſoftneſs ſhe, and ſweet attractive grace [1].

Whether the preſent age exhibits as many inſtances of ſu-
perior excellence as the preceding, is beyond my abilities to
determine: that it is tarniſhed more than ſome others with
one vice at leaſt, is generally confeſſed; and it may, perhaps,
be a problem not unworthy of conſideration, whether the
general uſe of Tea may not gradually increaſe the diſpoſition.
For whatever tends to debilitate, ſeems for the moſt part to
augment corporeal ſenſibility. The ſame perſon, who in health

[1] Milton's Paradiſe Loſt.

does

does not ftart at the firing of a cannon, fhall be extremely difconcerted when funk by difeafe to the border of effeminacy, at the fudden opening of a door. Defire is not always proportioned to bodily ftrength : it may fometimes be ftrongeft when the corporeal ftrength is at the loweft ebb ; it is often found fo ; and therefore another reafon occurs, why the general ufe of Tea ought not to be confidered as the moft indifferent of all fubjects.

From what has been faid upon this fubject, it will probably be admitted, that children and very young perfons in general fhould be deterred from the ufe of this infufion. It weakens their ftomachs, impairs the digeftive powers, and favours the generation of many difeafes. We feldom perceive the rudiments of fcrophulous difeafes fo often any where as in the weak feeble offspring of the inhabitants of towns, and whofe breakfaft and fupper often confift of the weak runnings of ordinary Tea, with its ufual appurtenances. It ought by no means to be the common diet of boarding-fchools ; if it be allowed fometimes as a treat, the children fhould at the fame time be informed, that the conftant ufe of it would be injurious to their health, ftrength, and conftitution in general.

N 2   SECTION

SECTION XIV.

THUS far I have chiefly endeavoured to trace the effects of Tea as a part of our diet. In medicine it has at prefent but very little reputation amongft us. It is even fcarcely ever recommended as a part of the furniture of a fick chamber; it is feldom mentioned even as a gentle diaphoretic: in cafes, however, where it is neceffary to dilute and relax, to promote the thinner fecretions, it promifes at leaft as much advantage as moft other infufions. For, befides its other effects, it feems to contain fomething fedative in its compofition (SECT. II. EXP. III. IV.), not altogether unlike an opiate. Like this clafs of medicines, it mitigates uneafinefs, perhaps more than any other merely aqueous infufion: and, like very fmall dofes of opium, it fometimes prevents reft, and gives a temporary flutter to the fpirits.

Where, therefore, large quantities of the infufion muft be taken, to produce or fupport a confiderable diaphorefis, a decoction of Tea, or a ftrong infufion, may be adminiftered with great propriety, particularly in inflammatory complaints; the fedative power of Tea, affifted by the diluting effects of warm water, generally producing a diaphorefis, without ftimulating the fyftem. The Chinefe moft commonly give it as a medicine in decoction, in a variety of difeafes; but if the infufion were

drawn

drawn from a large proportion of fine Tea, and foon poured off, that the fineft part may be procured, and drank warm, it would feem preferable as an attenuant and relaxant.

I have more than once given fine green Tea in fubftance with fome diluting vehicle, and obferved the fame effects nearly as are produced from taking the infufion. Thirty grains of this kind of Tea powdered, taken three or four times at as many hours interval, generally relaxes the folids, diminifhes heat and reftleffnefs, and induces perfpiration. Such a dofe as produces a flight naufea, which this quantity ufually does, more certainly induces a perfpiration, and a mitigation of the fymptoms accompanying inflammatory complaints. If this dofe be doubled, the naufea and ficknefs will be increafed, and a difagreeable fenfation or load is felt for fome time about the region of the ftomach, which ufually goes off with a laxative ftool.

SECTION

SECTION XV.

It is said that in Japan and China the stone is a very unusual distemper, and the natives suppose that Tea has the quality to prevent it [1]. So far as it softens and meliorates the water, which is very bad, it may certainly be of use [2]. We may also observe here, that every solvent is capable of taking up a limited quantity only of the solvend, and, when fully saturated with it, is incapable of suspending it long; hence it is plain, that the quantity of the stony matter carried off must be greater when the urine is increased in quantity, and has not been too long retained in the bladder : and therefore, as Tea is a diuretic, it may in this view prove lithonthriptic.

Tea, we have already observed, contains an astringent antiseptic quality (SECT. I. EXP. I, II.) It likewise possesses no inconsiderable degree of bitterness; and, as the uvæ ursi, and other bitters, have mitigated severe paroxysms of the stone, may not Tea prove serviceable also by its antacid quality ?

[1] Vid. Alex. Rhod. Sommaire, &c. J. N. Pechlin. Obf. xxvii. de Remed. Arthr. Prophylact. p. 276. Baglivius in doloribus calculosis et podagricis eam specialiter commendavit, p. 117. Vogel. Mat. Med. Thee Folia. Sir G. Staunton, Vol. II. p. 68, 69.

[2] By long boiling, water is certainly freed from some of the earthy and saline substances it may contain, and thereby rendered considerably softer; but it is by no means altered in these respects by infusing with Tea. See Percival's Experiments and Observations on Water, p. 27 et 33.

It

It is an obfervation I have often had occafion to make, that people, after violent exercife, or coming off a journey much fatigued, and affected with a fenfe of general uneafinefs, attended with thirft and great heat, by drinking a few cups of warm Tea, have generally experienced immediate refrefhment. It alfo proves a grateful diluent, and agreeable fedative, after a full meal, when the ftomach is oppreffed, the head pained, and the pulfe beats high[1]; hence the Poet fays,

> " The Mufe's friend, Tea, does our fancy aid,
> " Reprefs thofe vapours which the head invade,
> " And keeps that palace of the foul ferene,
> " Fit on her birth-day to falute a queen."
>
> WALLER.

[1] This is particularly remarked, as one of the good effects of Tea, by De Blegny, who wrote in 1680, which he probably copied from Alex. Rhod Sommaire des divers Voyages, &c. printed in 1653. See alfo Chamberlayn on Coffee, Tea, and Chocolate, p. 40. Le Compte's Memoirs and Obfervations, p. 227. Home's Principia Medicinæ, p. 5. Cheynæi Tractatus, p. 89. Percival's Experimental Effays, p. 130. Tiffot on the Difeafes of Literary and Sedentary perfons, p. 145, & feq. Dr. Kirkpatrick, in his notes upon this Work, relates the cafe of a Lawyer, who had been troubled for fome time with the gravel and ftone, and taken many medicines in vain; till at length he refolved to try the effects of Tea, an account of which is given by himfelf in the following words. " I had never ufed myfelf to " Tea, fo that the drink was new to me. I took a quarter of an ounce of fine bohea " Tea, and, pouring a quantity of boiling water upon it, fuffered the infufion to " ftand till it grew cold. I then poured it off clear, and drank three cups of it in " the morning, at the diftance of about an hour between each, two cups fafting, " one after breakfaft, and a fourth two hours after dinner. The firft day, the only " effect produced was a more plentiful difcharge of urine: but the fecond day I " voided in the morning twelve large fragments, a nucleus of the fize of a fmall pea, " with fome gravel; and what gave me more fatisfaction was, that the ufe of the " Tea kept my body open as in perfect health."

SECTION

SECTION XVI.

I SHALL finifh thefe remarks with fome reflections on this herb, confidered in another light.

As luxury of every kind has augmented in proportion to the increafe of foreign fuperfluities, it has contributed more or lefs its fhare towards the production of thofe low nervous difeafes, which are now fo frequent. Amongft thefe caufes, excefs in fpirituous liquors is one of the moft confiderable ; but the firft rife of this pernicious cuftom is often owing to the weaknefs and debility of the fyftem, brought on by the daily habit of drinking Tea [1] ; the trembling hand feeks a temporary relief in fome cordial, in order to refrefh and excite again the enfeebled fyftem ; whereby fuch almoft by neceflity fall into a habit of intemperance, and frequently intail upon their offspring a variety of diftempers, which otherwife probably would not have occurred.

Another bad confequence refulting from the univerfal cuftom of Tea-drinking, particularly affects the poor labouring people, whofe daily earnings are fcanty enough to procure them the neceffary conveniences of life, and wholefome diet. Many

[1] See Percival's Experimental Effays, p. 126. Duncan, in his Avis Salutaire, takes occafion to be merry upon the ufe and influence of Tea and hot liquors ; whilft he would not deprive voluptuous perfons of their idol, he would prevent it from burning its adorers, as *Moloch* did. Methufelah, he obferves, who lived near 1000 years, was a water-drinker ; but, fince the time of Noah, the firft wine-drinker, the life of man is contracted, and difeafes augmented.

of

of thefe, too defirous of vying with their fuperiors, and imitating their luxuries, throw away their little earnings upon this foreign herb, and are thereby inconfiderately deprived of the means to purchafe proper wholefome food for themfelves and their families. In the words of Perfius we may here juftly exclaim,

O curas hominum quantum eft in rebus inane !

I have known feveral miferable families thus infatuated, their emaciated children labouring under various ailments depending upon indigeftion, debility, and relaxation. Some at length have been fo enfeebled, that their limbs have become diftorted, their countenance pale, and a marafmus has clofed the tragedy[1].

Thefe effects are not to be attributed fo much to the peculiar properties of this coftly vegetable, as to the want of proper food, which the expence of the former deprived thefe poor people from procuring. I knew a family, confifting of a mother and feveral children, whofe fondnefs for Tea was fo great, that three times a day, as often as their meals, which generally confifted of the fame articles, they regularly fent for Tea and fugar, with a morfel of bread to fupport nature ; by which practice, and the want of a due quantity of nutritious food, they grew more enfeebled ; thin, emaciated habits and weak conftitutions characterifed this

[1] See Dr. Walker's excellent Remarks, in Memoirs of the Medical Society, Vol. II. p. 43.

diftreffed

diftrefied family, till fome of the children were removed from
this baneful nurfery, by which they acquired tolerable health.

My valuable friend, Dr. Walker, of Leeds, in Yorkfhire,
has noticed, in feveral parts of that extenfive and commercial
county, and particularly in Leeds; that, " fince the more
plentiful introduction of Tea into the families of the induftrious
poor, by the late reduction of its price, the Atrophia Lactan-
tium, or Tabes Nutricum, a fpecies of decline, has made an
unufually rapid progrefs. The difficulty with which animal food
is procured by the lower ranks of fociety, in quantity fufficient
for daily nutriment, has led many of them to fubftitute, in the
place of more wholefome provifions, a cheap infufion of this
foreign vegetable, whofe grateful flavour (and perhaps narcotic
quality, which it poffeffes in a fmall degree in common with
moft other ever-greens) is found to create an appetite for
itfelf, in preference to all other kinds of aliment that the fcanty
income of poverty allows thefe deluded objects to procure ;
though I am forry to have occafion to add, that the lowering
effects of tea-drinking lead too many of thefe to feek relief
from fpirits, and other pernicious cordials, at the expence of
health, and the fure confequences of penury and want.

" As this change, in the article of diet, has been very generally made, efpecially by the females, and the younger
branches of the families of the manufacturing poor, their
conftitutions have been rendered much lefs able to bear
evacuations of any fort, and particularly that of lactation. I
may, with great truth, aver, that more than two hundred
patients of this denomination have, within the laft two years,
come

come under my notice : upon their application for relief, and
the confequent enquiry which I have been led to make
refpecting the nature of their diet, their almoft invariable
reply has been, that they have chiefly depended upon Tea for
their fupport, at the fame time that they were permitting an
apparently healthy child to draw the whole of its nourifhment
from them.

" That it is debility, and an impoverifhed ftate of the whole
fyftem, arifing from a deficiency in the due fupply of proper
and fufficiently nutritious aliment, at a time when the conftitution particularly requires it, in confequence of the continual
wafte which the mother fuftains from the fuckling of her infant,
which lay the foundation of this difeafe, and that the lungs are
but fecondarily or fymptomatically affected, is clearly evinced
from an attention to the fymptoms.

" The patient firft complains of languor, and general weaknefs ;
lofs of appetite ; fatigue after exercife, though it be of the
gentleft kind ; wearifome pains in the back and limbs ; foon
after which, fymptoms of general atrophy come on ; the face,
in particular, grows thin, and is marked by a certain delicacy
of complexion ; palenefs about the nofe ; but with a fmall
degree of fettled rednefs in the cheeks. In a fhort time, if
the patient ftill continues to give fuck, fhe is feized with
tranfitory ftitches in the fides, under the fternum, or in fome
other part of the thorax ; accompanied with a fhort dry cough,
and flight dyfpnæa, upon any mufcular exertion ; the pulfe alfo
becomes frequent, but feldom fo hard as in the inflammatory
ftate of the genuine phthifis pulmonalis ; morning fweats next

O 2   make

make their appearance; abſceſſes and ulcers are often formed in the lungs; pus mixed with mucus is expectorated; the general weakneſs increaſes; the emaciated patient is unable to ſupport an erect poſture; and at laſt dies literally exhauſted."

An ingenious author obſerves, that as much ſuperfluous money is expended on Tea and Sugar in this kingdom, as would maintain four millions more of ſubjects in bread [1]. And the author of the Farmer's Letters calculates, that the entertainment of ſipping Tea coſts the poor each time as follows:

				$d.$
The tea	—	—	—	$\frac{1}{2}$
The ſugar	—	—	—	$\frac{1}{4}$
The butter	—	—	1	
The fuel and wear of the Tea equipage		—		$\frac{1}{4}$
				$2\frac{1}{2}$

When Tea is uſed twice a day, the annual expence amounts to 7l. 12s. a head. And the ſame judicious writer eſtimates the bread, neceſſary for a labourer's family of five perſons, at 14l. 15s. 9d. per annum [2]. By which it appears, that the yearly expence of Tea, Sugar, &c. for two perſons, exceeds that of the neceſſary article of bread, ſufficient for a family of five perſons.

[1] Eſſays on Huſbandry, p. 166.   [2] Vol. I. p. 202. and 299.

It

It appears alfo, from a moderate calculation, that twenty-one millions of pounds of Tea [1] are annually imported into England. In the beginning of the prefent century the annual public fales by the Eaft-India Company did not much exceed 50,000 pounds weight, independently of what little might be clandeftinely imported. The Company's annual fales about this time, 1797, approach to twenty millions of pounds; being an increafe of four hundred fold in lefs than 100 years, and anfwers to the rate of more than a pound weight each in the courfe of the year, for the individuals of all ranks, fexes, and ages, throughout the Britifh dominions in Europe and America [2].

Since the year 1797, it is probable, that the import of Tea has increafed in a much greater ratio; for the Eaft-India Company, at their fale in September 1798, put up 1,300,000 pounds of bohea; 3,500,000 pounds of congou and campoi; 400,000 pounds of fouchong and pekoe; 600,000 pounds of finglo and twankay; 400,000 of hyfon; hyfon fkin 100,000; making, in the whole, 6,300,000 pounds, the quantity fold in the autumnal quarterly fale: and it may be prefumed, from the table annexed, (p. 1. Section IV.) and other documents, that at leaft 30,000,000 of pounds are annually imported into Europe and America!

[1] If we include the quantity fmuggled into this kingdom, the confumption might be calculated at half a million more.

[2] Compare Sir George Staunton's Embaffy, vol. I. p. 22.

F I N I S.

DIRECTIONS FOR THE PLATES.

―――――――――――――――

ERRATUM.

P. 41. l. 10. for *than that* read *as*.